No. 2677
$19.95

BASIC
ROOF
FRAMING

BENJAMIN BARNOW

TAB BOOKS Inc.
Blue Ridge Summit, PA 17214

FIRST EDITION
FIRST PRINTING

Library of Congress Cataloging in Publication Data

Barnow, Benjamin.
Basic roof framing.

Includes index.
1. Roofs. 2. Framing (Building) I. Title.
TH2393.B37   1986   694'.2   86-1917
ISBN 0-8306-0677-7
ISBN 0-8306-2677-8 (pbk.)

Front cover photographs:
Top photograph courtesy of Future Homes 5905 Snider Road, Saint Thomas, PA 17252. Bottom photograph courtesy of Berry and Homer Photographics, 1210 Race Street, Philadelphia, PA 19107

Contents

Above all I wish to thank my wife Flora. Without her help, patience, and understanding this book would not have materialized.

Introduction

Basic Roof Framing was written with a twofold purpose in mind. First it is a book for the professional carpenter who did not receive a structured training in the principles of roof construction. Second it is for those who do not make their livelihood as carpenters but enjoy working with wood and want to expand their knowledge and skill in this field. It is a balanced text between theory and practical application in one of the more complicated phases of house construction.

The first few chapters cover tools, safety, materials, math, and the use of the calculator through the principles and use of the framing square. Then it's on to the types of roofs, roof framing terms, and how they are related. Then comes the actual calculation, layout, cutting, and erection of the common rafter as well as sheathing and trimming of the roof. The explanations are as simple and uncomplicated as possible.

Included is a unit on shed roofs for those who might have the desire and need to go on to a more advanced type of work. I am confident that, with a fair amount of skill and patience, those who have the desire can build a well-proportioned, well-constructed, and safe roof for any addition, shed, or a complete house.

I do suggest that you read the entire book and try to visualize each step involved. All good craftsmen develop the ability to visualize a project they want to accomplish before construction begins. Without this ability, it is most difficult, and at times almost impossible, to accomplish the desired task.

The appendices include general construction terms used in this book that are not explained in the text, a list of the most frequently used formulas for roof framing, and a number of charts and tables to help speed and make your job easier.

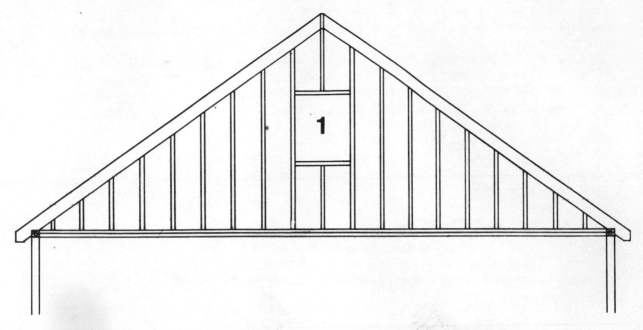

Preliminaries

Objective. This chapter will acquaint you with the preliminary information needed to construct a roof. The skills, tools, and materials needed to build the roof safely and efficiently are described.

Upon completion of this chapter, you should be familiar with the following information.

- ☐ The skills needed to build a roof.
- ☐ General safety in construction.
- ☐ Safe use of ladders.
- ☐ Safety for power-tool users.
- ☐ Electrical safety for power tools.
- ☐ Planks and scaffolds.
- ☐ Hand tools needed for roof framing.
- ☐ Power tools for roof framers.
- ☐ Choosing the proper materials for framing a roof.
- ☐ Why wood is used as a framing material.
- ☐ How lumber is classified and graded.
- ☐ How plywood is classified and graded.
- ☐ Fasteners used in roof framing.

BEFORE YOU BEGIN

It is not much more difficult to build a roof than to erect the walls and floors. Yes, it does require some math skills but none that the average reader should have trouble mastering. A math refresher is given in Chapter 2; these skills will be needed to lay out rafters accurately.

The practical carpentry skills required for roof framing are no greater than those encountered in erecting the walls. The greatest difference is that erecting a roof takes place off the ground. This means that ladders must be used, and at times scaffolding and planks will be required, but mostly you will need the ability and confidence to climb and work above ground level. To erect a roof, your crew should consist of at least two persons. Using four persons, if they are available, will make the job easier and move along much quicker.

SAFETY ON THE JOB

Safety is of primary importance in construction. You must be conscious of safety at all times. Most accidents can be prevented by acquainting yourself with safety precautions and by developing the correct attitude towards them.

General Safety

☐ Keep the job site clean.
☐ Watch your footing; walk do not run.
☐ Stack all material neatly.
☐ Help prevent fires; dispose of all rubbish.
☐ Protruding nails should be removed or bent down.
☐ Do not drop tools or materials on those working below you. Be careful.
☐ Above all, concentrate on what you are doing.

Personal Safety

☐ Clothing should suit the working conditions.
☐ All clothing should fit properly.
☐ Sturdy work shoes with nonslip soles are best.
☐ Any loose fitting clothing should be tied or fastened properly.
☐ Long hair should be contained with a cap or hairnet.
☐ No jewelry, rings, watches, chains, or bracelets should be worn.

Ladder Safety

In the process of erecting the roof, a certain amount of climbing—including the use of ladders—will be required. Keeping the following precautions in mind and using common sense, you should be able to build a roof with a minimum of wasted effort—and do it safely.

Ladders are made of wood, aluminium, and fiberglass. Wood and aluminium are more popular because of cost (wood) and light weight (aluminium). If you plan to use your ladder in the vicinity of electrical lines, choose wood or fiberglass. Never touch an electrical line with any type of ladder.

Fig. 1-1. Extension ladder. Courtesy of Werner Ladders.

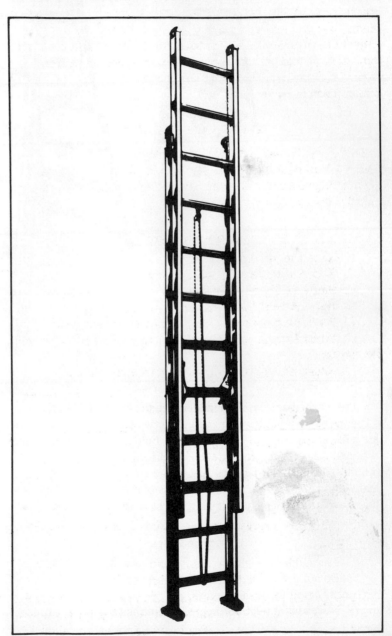

Two of the basic types of ladders are the extension ladder (Fig. 1-1) and the step ladder (Fig. 1-2). The two-section extension ladder will extend from 13 feet to 43 feet. Folding step ladders come in height sizes from 3 feet to 20 feet. Prices vary with the quality and classifications of ladders that range from homeowner use to heavy-duty, commercial use. Regardless of classification, most ladders have a seal of safety stating the safe working loads.

Fig. 1-2. Step ladder. Courtesy of Werner Ladders.

Using Ladders

☐ The ladder should be sturdy and in good condition.

☐ Do not use the ladder if it wobbles or if it has cracks, missing rungs, or other missing parts.

☐ Never paint a ladder because paint has a tendency to hide defects.

☐ Do not overreach. Choose a ladder of the proper length to accommodate the height at which you are working.

☐ Move the ladder as needed so that reaching over the side is not necessary.

☐ Place all ladders on firm, level ground. If necessary, block and adjust the ladder with solid boards at the foot of the ladder.

☐ Tie off the base and top of the ladder to keep it from slipping.

☐ Keep rungs clean and dry.

☐ Always face the ladder when climbing or descending.

☐ Hold on with two hands.

☐ Never use the top step of a ladder as a tread.

☐ On an extension ladder, make sure it extends at least 3 feet above the top edge of the roof. See Fig. 1-3.

☐ Place the bottom of the ladder out in a horizontal direction one-quarter of the distance that the ladder is extended. An angle of 70 to 75 degrees is appropriate. See Fig. 1-3.

☐ Make sure step ladders are fully open.

Fig. 1-3. Proper placement of a ladder against a building.

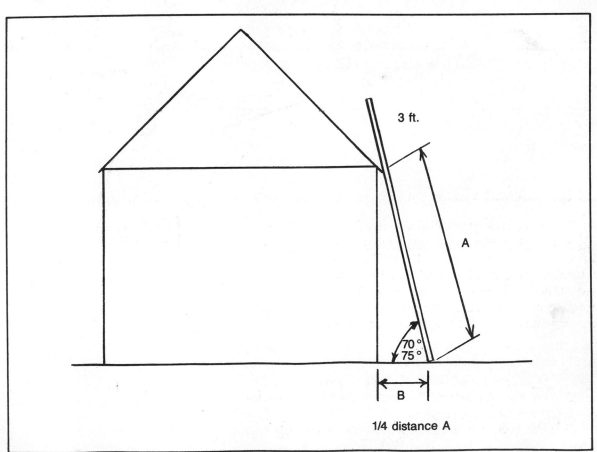

3 ft.

A

70°
75°

B

1/4 distance A

☐ Make sure all locking devices are properly engaged.
☐ Above all take your time.

Scaffold Safety

A scaffold is a temporary platform where you can work and stand safely. Materials, tools, and equipment can be stored on a scaffold to be used as needed. Scaffolding can be erected to most any height when properly engineered. There are two types of scaffold available: wood scaffolding that is constructed on the job and metal scaffolding that can be bought or rented as needed. Wood scaffold is usually used only when small sections are needed. Because manufactured scaffolding is safe, portable, economical, easy to erect, and can be used over again, it is a popular choice for craftsmen.

For framing a roof, scaffolding is not always necessary, but it does make the job of trimming and finishing the structure easier, quicker, and safer. It pays to consider scaffolding for framing a roof. See Figs. 1-4A and 1-4B.

Using Scaffolding

☐ Scaffolding should be placed on firm and level ground.

☐ If conditions require, adjustable legs should be used to level and plumb the scaffold. The legs should be placed on wide, heavy planks for adequate support.

☐ Freestanding scaffolding should be securely fastened to the building.

☐ Make sure all locking devices are engaged.

☐ Always use a ladder or stairs to reach the scaffold platform.

☐ Never use a ladder or any other device to stand on the scaffold platform.

☐ Make sure all lumber used for planks are sound and of good quality.

☐ All planks should overlap at least 12 inches, extend at least 6 inches each side of the supports, and are fastened so that they cannot slip.

Wall Brackets

In addition to scaffolding, wall brackets are used by carpenters when working above ground. Used in place of scaffolding, wall brackets are fastened to the building with nails or bolts. They are usually placed 6 to 8 feet apart. A platform is created with planks placed across them. This gives the carpenter a temporary platform from which to work in safety and comfort. See Fig. 1-5.

Fig. 1-4A. Freestanding metal commercial scaffolding. Courtesy of Werner Ladders.

Fig. 1-4B. Scaffolding in place.

Fig. 1-5. Wall brackets in place on
a building.

Using Wall Brackets

☐ All fasteners must be secured into solid wood. The fasteners
must penetrate the sheathing and be anchored into studs or adequate blocking must be placed behind the sheathing.

☐ All precautions recommended for scaffold planks should be
followed when using wall brackets.

☐ When using wall brackets, it is recommended that 20 (D)
penny nails be used. Upon removal of the brackets, the nails are
then driven flush.

TOOLS USED IN ROOF FRAMING

Roof framing can be accomplished with a limited number of basic hand and power tools. The professional carpenter chooses his tools wisely. In choosing a tool, pick a good-quality, well-constructed tool. Buy the best tools you can afford. Take proper care of your tools and you will reap benefits during the years of use you will have from them.

Hand Tools

Hammers. Hammers come in many sizes, shapes, weights, and handle sizes. The usual weights are 16 to 32 ounces. For roof framing, a hammer weight of 22 or 24 ounces is recommended. Some carpenters, however, prefer a 20-ounce hammer and some feel comfortable with a 28-ounce hammer. The shape of hammer heads are curved claw or straight claw that is also called a ripping hammer. Most carpenters prefer the ripping hammer for framing.

Hammer handles are made of wood, steel, and fiberglass. Some of the materials used for handles are left natural while others are covered with rubber or vinyl. Wood handles are shaped round or octagon. The choice of handle shape is yours. In buying a hammer, compare the different types, swing them, feel the weight and balance, then make the decision as to which one feels best to you. A good hammer will be well constructed and give many years of useful service if it is not abused. See Figs. 1-6, 1-7, and 1-8.

Measuring Tapes. A 25-foot tape with a 1-inch-wide blade is best for short measurements. Use a 100-foot steel tape for long measurements. See Fig. 1-9.

Squares for Framing. The steel framing square (as described in detail in Chapter 4) is a most important tool for the roof framer. See Fig. 1-10.

Fig. 1-6. Curved claw, steel-handle framing hammer. Courtesy of Stanley Tools.

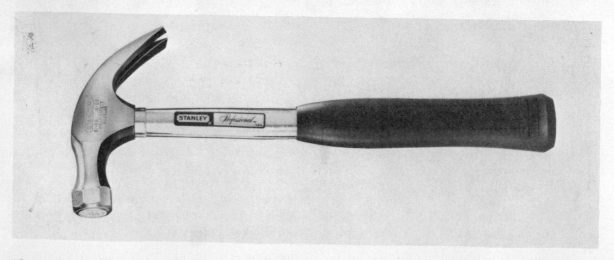

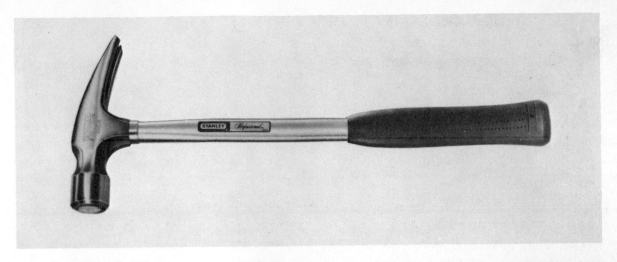

Fig. 1-7. Straight claw (ripping) hammer with a steel handle. Courtesy Stanley Tools.

The try square (Fig. 1-11) or the adjustable square (Fig. 1-12) are used for marking boards to make square cuts. The adjustable square has more uses than the try square, but it can lose its accuracy quicker than the try square.

The sliding T bevel is another square used for transferring angles or marking a number of repeating angles. See Fig. 1-13.

Levels. Levels are manufactured in 6-inch to 6-foot sizes. A good double-vial, 48-inch wood level (Fig. 1-14) or an aluminium level (Fig. 1-15) will suffice for most all framing work. If you choose a wooden level, buy one that is metal bound. The metal edge helps maintain accuracy. As with any other good tool, take care of your level, and don't drop it.

Plumb Bob. A plumb bob is a tool that the carpenter uses to place an object in a perfectly vertical position. It is a pointed

Fig. 1-8. Straight claw (ripping) hammer wood handle. Courtesy Stanley Tools.

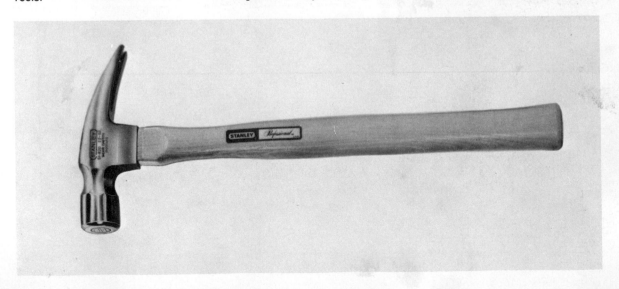

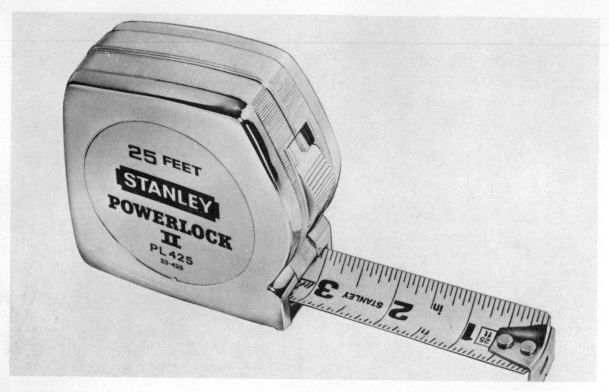

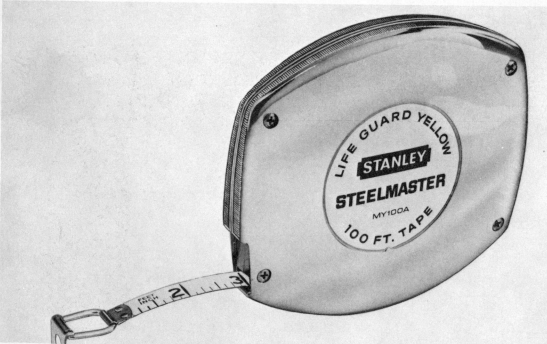

Fig. 1-9. A 25-foot power return tape and a 100-foot steel tape. Courtesy Stanley Tools.

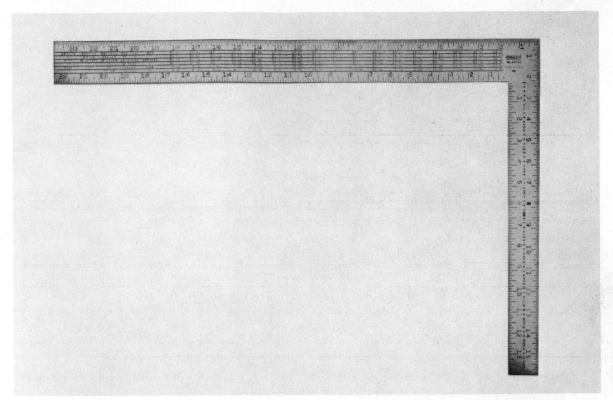

Fig. 1-10. Steel roof-framing square.

Fig. 1-11. Professional hardwood-handle try square. Courtesy Stanley Tools.

13

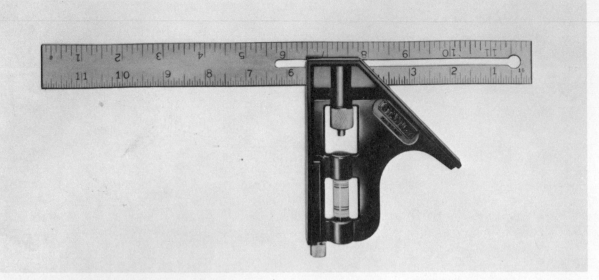

metal weight that when suspended from a string will hang absolutely vertical (plumb). See Fig. 1-16.

Chalk Box and Line. Purchase a chalk box that will hold 100 feet of line plus chalk. Some boxes are made so that they can double as a plumb bob if necessary. See Fig. 1-17.

Hand Saw. A standard 26-inch, 8-point crosscut saw of pro-

Fig. 1-12. Adjustable combination square. Courtesy Stanley Tools.

Fig. 1-13. A sliding T bevel. Courtesy Stanley Tools.

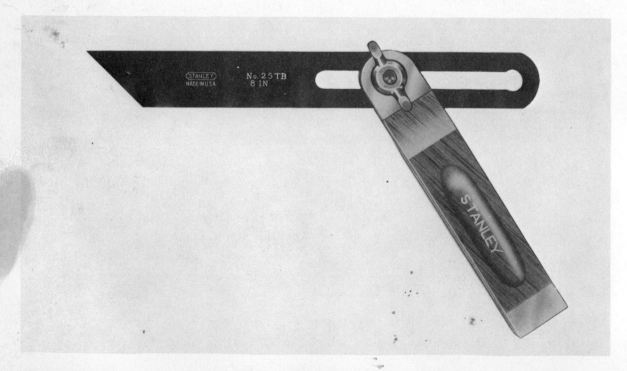

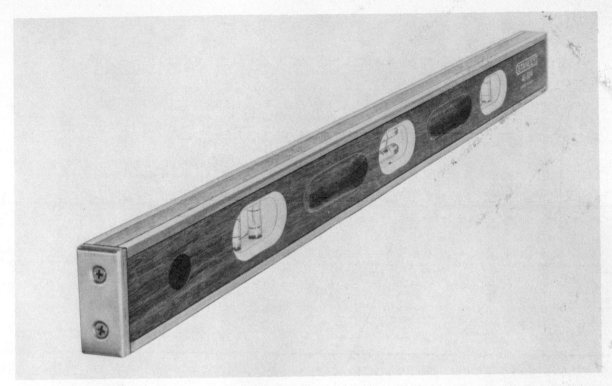

Fig. 1-14. Professional brassbound mahogany mason's level.

Fig. 1-15. Professional aluminium level. Courtesy Stanley Tools,.

fessional quality is your best choice for a hand saw. A quality saw will keep its sharp edge for a longer period of time. The secret of effortless hand sawing is to keep it sharp. See Fig. 1-18.

Nail Puller. Sometimes called a cat's-paw, a nail puller will help in correcting those inevitable errors that will occur. This tool will remove nails that have been set with less chance of destroying the piece of lumber involved. See Fig. 1-19.

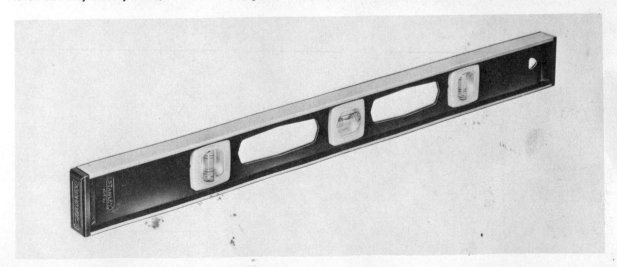

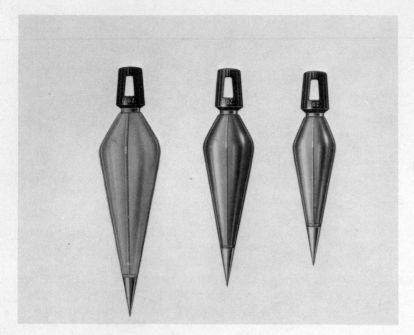

Fig. 1-16. Plumb bobs (5-, 8-, and 12-ounce). Courtesy Stanley Tools.

Fig. 1-17. A professional chalk box, with 100 feet of line. Courtesy Stanley Tools.

Fig. 1-18. A professional 26-inch handsaw. Courtesy Stanley Tools.

Hatchet. A hatchet is very handy when rough cutting has to be done. A hatchet will handle the job much quicker than some other cutting tools. With practice and care, some fine cutting and trimming can be done with a hatchet. See Fig. 1-20.

Utility Knife. A utility knife is a very handy tool for all those odd cutting jobs. You can also use your knife to keep your pencil sharp (a must) for accurate roof framing. See Fig. 1-21.

Tool and Nail Bag. You won't be able to work without a tool or nail bag. It holds an assortment of nails plus the most-often used tools for carpentry work. In the interest of economy, a cloth nail bag (which many lumberyards supply free) and a hammer holder will get you through the job.

Ripping Chisel. A ripping chisel is a most handy tool to remove (without damage) a piece of stock that has been put in the wrong place. Notice the beveled nail slot (Fig. 1-22) to make the removal of nails easier.

Safety Goggles. Goggles are a must when performing any striking operation or cutting with any power tool. See Fig. 1-23.

Fig. 1-19. Nail puller with 90-degree offset claw. Courtesy Stanley Tools.

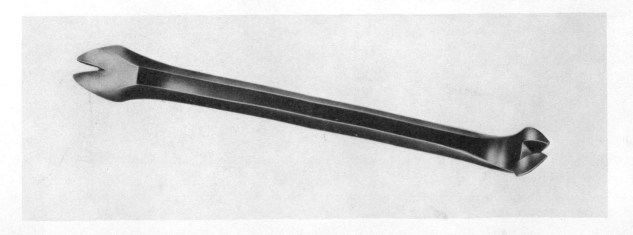

17

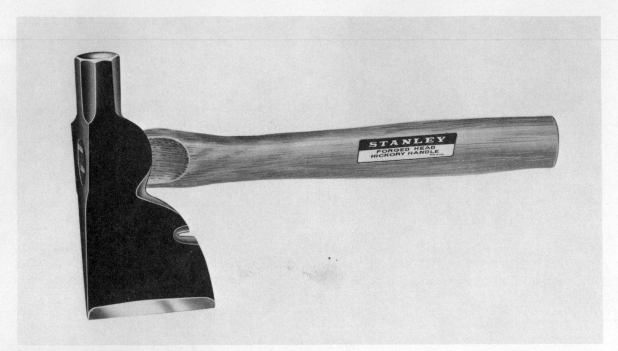

Fig. 1-20. A wood-handled half hatchet. Courtesy Stanley Tools.

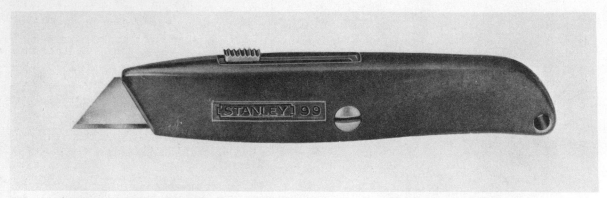

Fig. 1-21. A retractable-blade utility knife. Courtesy Stanley Tools.

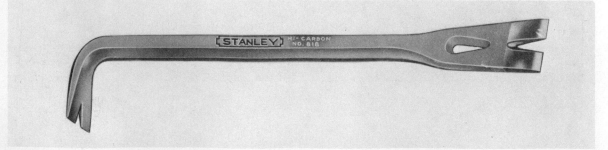

Fig. 1-22. Beveled ripping chisel. Courtesy Stanley Tools.

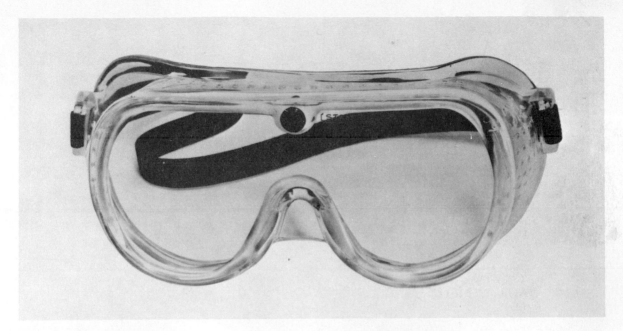

Fig. 1-23. Safety goggles. Courtesy of Stanley Tools.

Power Tools

Circular Saw. A portable circular power saw is the only power tool needed to do the average roof-framing job. A 7 1/4-inch blade is one of the more popular sizes. Bear in mind that as the blade size is increased so does the weight of the saw. Check the horsepower, current draw, and type of bearings used; these features all help control the quality of the saw. A low-priced saw might not last through the job. Much depends on the size and number of rafters you will have to cut. A saw (Fig. 1-24) that is rated as the top of the consumer line will perform very well because this is about the same quality as a low-end professional contractors' saw.

Reciprocating Saw. The reciprocating saw (Fig. 1-25) is very handy for those cuts that cannot be made with a circular saw. Due to the reciprocating action of the blade, it can reach places and make cuts that would be impossible to do with the standard circular saw. It is a most handy power tool to have but it is not an absolute must.

Using Power Tools Safely

General safety rules for most power tools are the same. I suggest that you read the owners manual that comes with the tool until you understand the correct way to use the tool in an efficient and safe manner.

□ Always disconnect the power source to change the blade or make any adjustments.

□ Make sure the blade is installed properly and the arbor nut is tight.

☐ Allow the saw to come to full speed before starting to cut.

☐ Be careful that the saw does not kick back (this does happen frequently). Never stand directly behind the saw; always stand to one side of it.

☐ Keep your fingers clear of the cutting blade.

☐ Do not remove the guard.

☐ Set the saw blade so that it is about 1/8 of an inch deeper

Fig. 1-24. Circular saw. Courtesy of Black and Decker Tools.

Fig. 1-25. Reciprocating saw. Courtesy of Black and Decker Tools.

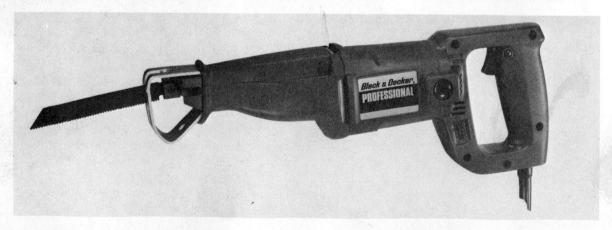

than the stock being cut. It is safer and the saw operates more efficiently.

☐ Wait until the saw comes to a complete stop before setting it down.

☐ Above all, use the proper blade, as recommended by the manufacturer, and keep it sharp.

Electrical Safety For Power Tools

The improper use of electrical power tools can lead to serious injury or even death. You must maintain and use electric power tools properly. Grounding a power tool is most important (unless insulated power tools are being used). The National Electric Code (NEC) requires that all electrical circuits be grounded. Because this ruling was not always in effect, unfamiliar electrical circuits must be checked to make sure they are grounded before any tools are plugged in. This can be accomplished very easily with the aid of an inexpensive voltage tester. Either a store-bought, neon-type tester or one made up of a pigtail socket and light bulb can be used. To ground a tool properly, a third wire (ground wire) is used. This wire is attached to the tool case and electrical system ground by the use of the third (usually round) prong on the power cord to the wall outlet.

In recent years, electric power tools with double and triple insulation have become standard products with all power tool manufacturers. The metal case has been eliminated and replaced with plastic. This prevents the user from coming in contact with any electric current if the insulation on the inside of the tool breaks down. These tools come with two-wire power cords. There is no need for the extra ground wire. Such tools are safe to use without external grounding, provided the manufacturer's directions are followed. It is a good policy to have any electric power tools checked regularly and to observe the following precautions.

☐ Never use power tools if there is any chance of coming in contact with water, this includes rain, puddles, moisture and even excessive body perspiration.

☐ On three-wire extension cords never remove the ground prong.

☐ If the wall outlet does not have a three wire socket use a three-to-two prong adapter. Check and make sure the system is grounded.

☐ Always disconnect the power plug when the work is completed.

☐ Use of the proper size power cord will increase tool efficiency and life.

PROGRESS CHECK

1. When working in the vicinity of electrical lines _____ ladders should be used.

2. It is best to paint a ladder to preserve it. T — F.

3. An extension ladder should extend at least 3 feet above the roof. T — F.

4. The base of the ladder should extend out in a horizontal direction _____ the distance it is extended.

5. In place of ladders a temporary platform called _____ is often used.

6. To reach the scaffold platform, a _____ should be used.

7. Wall brackets are fastened to the building with _____ or _____, which penetrate the sheathing into a solid framing member.

8. If nails are used to fasten wall brackets, a _____ _____ nail is the minimum size.

9. A large number of special tools are needed to frame a roof. T — F.

10. A _____ _____ _____ is one of the most important tools used in roof framing.

11. A circular saw with a _____-inch blade is one of the most popular saws for general carpentry work.

12. Always _____ the power source before making any adjustments on an electric power tool.

13. Allow a saw to come to _____ _____ before starting to cut.

14. The teeth of the saw blade should be set about _____ of an inch deeper than the stock being cut.

15. If insulated power tools are used, grounding is not necessary. T — F.

16. A _____ _____ is used to check electrical circuits.

17. It is best to check any unfamiliar electrical circuits for proper grounding before using them. T — F.

18. For proper grounding, extension cords should contain _____ wires.

19. If the wall outlet is not grounded, a _____ _____ hooked up properly may be used.

20. Use of the proper size power cord has no bearing on the life of a power tool. T — F.

WOOD AS A FRAMING MATERIAL

Wood is one of the basic and most widely used building materials. It is readily available in abundant quantities. With proper forestry management, wood replaces itself approximately every 25 years.

Wood is easily cut and is light in weight as compared to other building materials. It is strong yet flexible, can be transported easily, it can be handled on a job site without expensive equipment, and it is easily fastened with a variety of inexpensive fasteners. Wood can be formed into decorative patterns, combines well with other materials, and can be finished in many different ways.

Even though wood has many advantages—such as natural insulating qualities, resistance to acids, saltwater, and other corrosive agents—it also has some disadvantages. It can be destroyed or weakened by some chewing and boring insects. If not properly cared for, it is vulnerable to disease and decay. Wood shrinks and expands with moisture content, can warp or crack, and is not always consistent in strength or hardness. In spite of these faults, it still is an ideal building material. With chemical treatments for decay and insects, proper grading for strength and quality, and proper handling on the job site, it is the number one choice of material for residential construction.

Wood Identified By Type

Wood is classified as hardwood or softwood. This method of identification does not actually indicate the softness or hardness of wood but identifies the two main types of trees from which lumber is manufactured.

Softwoods come from evergreen trees that are cone- or needle-bearing trees (they are called conifers). Some common trees in this category are pine, fir, hemlock, and spruce.

Hardwoods are sawn from broad leaf trees (called deciduous) which lose their leaves once every year, some popular hardwoods are oak, maple, birch, cherry and walnut.

Practically all construction lumber materials are softwoods. Hardwoods can be used for trim or special units in house building, but very rarely for framing.

Cutting Lumber From Logs

After the tree is felled, it is cut into logs of 12 feet to 40 feet in length and transported to the saw mill. Under the guidance of expert sawyers, it is cut into lumber. Methods of cutting vary with each tree in order to minimize waste and to get the strongest and best appearing boards possible from each log.

Moisture Content and Drying

After a board is cut, edge trimmed, and cut to width, it must be dried and planed to size. When a tree is first cut and sawn into lumber, the green wood has a moisture content of 35 to 300 percent of its weight as compared to dry wood. There is no shrinkage in lumber until the moisture content (abbreviated MC) drops below 28 percent. To be suitable for ordinary house construction, it has been found that an MC of about 16 to 19 percent is desired.

Air and Kiln Drying Lumber

To attain an MC of 16 percent, a seasoning method called air drying (a natural method) is used. The lumber is arranged in stacks, with thin crosspieces of wood called stickers between the layers, to allow a natural air flow. It remains outdoors in these stacks from one to three months. At that time it should have reached an MC of 19 percent. Air drying time required is related to the moisture content of the surrounding air.

Even though a 16 to 19 percent MC is acceptable for framing lumber, there are times when it is necessary to reach much lower moisture contents. Then a method called kiln drying (abbreviated KD) is used. The lumber is stacked (the same as for air drying) indoors in large sheds. The doors are closed and hot air is circulated through the building, removing the moisture from the lumber. The MC can then be brought down to the 6 or 7 percent level required for furniture, flooring, and interior woodwork.

The MC is checked by the use of meters that measure, with an accuracy range of plus or minus 1 percent, the electrical resistance in the lumber. Softwoods can be seasoned by the kiln drying method in two to 10 days. Depending on the size and species of lumber, hardwoods take 10 days to three weeks to season. Air drying (natural seasoning) can take up to two years according to the class of material and use to which it will be put. Green or unseasoned lumber has a moisture content of more than 19 percent. Dry or seasoned lumber has a moisture content of less than 19 percent.

GRADING LUMBER

Once the tree has been felled and cut into lumber, it is then graded. Grading is sorting and classifying wood in accordance to qualities, strengths, and defects. This is necessary so that the consumer can know the quality of lumber that he is buying. Lumber is graded in accordance with two sets of rules: hardwood lumber grading and softwood lumber grading. Because a relatively small amount of hardwood is used in residential basic roof construction, I will limit this discussion of lumber grading to softwoods.

Grading Softwood Lumber

The National Bureau of Standards has outlined a set of grading rules in cooperation with growers, lumber organizations, government agencies, distributors, and users of lumber. The standards of grading were developed by the major lumber associations in the United States. Portions of the quoted information and grading rules were taken from the following sources:

West Coast Lumber Inspection Bureau Standard Grading Rules
    Number 16, Revised 1984
P.O. Box 23145
Portland, Oregon 97233.

Southern Pine Inspection Bureau Grading Rules
4709 Scenic Highway
Pensacola, Florida 32505.

Classes of Lumber

Lumber is divided into three broad classes.

Yard Lumber. Yard lumber is timber that is cut into sizes, grades, and patterns for construction and general uses. Yard lumber includes common boards, wood siding, sheathing, flooring and dimension lumber (2×4s, 2×6s, etc.). It is used for posts, joists, roof rafters and timbers to support heavy loads.

Structural Lumber. Structural lumber is used where great strength is needed. Examples are joists beams, posts, and rafters used for roofs when required by engineering design. Structural lumber is graded for structural strength rather than appearance. The lumber is 2 inches and greater in thickness and 4 inches and greater in width. The word *structural* appears in the grade stamp.

Factory and Shop Lumber. Factory and shop lumber is intended for making doors, windows, and other millwork. It is for further manufacture in mill shops and wood manufacturing facilities. It is rarely used in house framing.

Grading of Dimension Lumber

Because lumber grading is a broad and detailed subject, I will limit my discussion to the information you will need to construct a roof and sheath it. Dimension lumber and plywood sheathing are the two common types of lumber used in roof framing.

Dimension lumber is limited to surfaced softwood lumber of nominal thickness (from 2 through 4 inches), and is designed for use as framing members such as joists, planks, rafters, studs, and small timbers.

Classification of Dimension Lumber

Dimension lumber is classified into two width categories and five use categories.

☐ Dimension lumber up to 4 inches wide is classified as *Structural Light Framing*, *Light Framing*, and *Studs*.

☐ Dimension lumber 5 inches and wider is classified as *Studs* and *Structural Joists and Planks*.

In addition, a single appearance grade is designed for special uses where strength with high appearance is needed.

In roof framing for residential construction, *Light Framing* and *Structural Joists and Planks* are the most common classifications used.

Light Framing

2-4 inches thick, 2-4 inches wide
Construction (Const) Grade
Standard (Stand) Grade
Utility (Util) Grade

Light Framing grades are designed to provide dimension lumber of good appearance at lower design levels for all those uses where higher design levels coupled with high appearance are not needed.

Structural Joists and Planks

Structural joist and plank grades are designed especially to fit engineering applications for lumber 5 inches and wider. There are four grades in this category—with *Select Structural* being the highest and Grade No. 3 being the lowest—and there is one single stud grade.

2-4 inches thick, 5 inches and wider
Graded Select Structural (Sel Str)
Grade No. 1
Grade No. 2
Grade No. 3

Studs

2-4 inches thick, 2-6 inches wide

This single stud grade comes in 10-foot lengths or shorter. Some of the species of trees covered under these rules are loblolly, long leaf, short leaf, pond, Virginia, and spruce.

The standards for the above grading information was supplied by the Southern Pine Inspection Bureau Grading rules.

West Coast Lumber Grading Rules

The grading rules for West Coast Lumber are the same as for Southern Pine, with some minor modifications to accommodate some species of lumber common to the West Coast. All grading rules were developed in accordance to the standards set by the National Bureau of Standards.

The categories for West Coast Dimension Lumber are *Structural Light Framing, Light Framing, Studs, Structural Joists and Planks*, and *Appearance Framing*.

Grades of Light Framing Lumber

There are four grades of light framing lumber.

Constrution Grade. Construction-grade lumber is recommended and widely used for general framing purposes. Pieces are of good appearance but are primarily graded for strength and serviceability.

Standard Grade. Standard-grade lumber is used for the same purposes as construction-grade lumber. Characteristics are limited to provide good strength and excellent serviceability.

Utility Grade. Utility-grade lumber is recommended and widely used where a combination of good strength and economical construction is desired for such purposes as studding, blocking, plates, bracing, and rafters.

Economy Grade. Economy-grade lumber is suitable for crating, bracing dunnage, and temporary construction.

Structural Joists and Planks

Structural joists and planks are graded to the same standards and requirements as Southern Pine.

Some of the species included under the West Coast Lumber Grading Rules are: Douglas fir, larch, western hemlock, Pacific fir, noble fir, California red fir, white fir, mountain hemlock, sitka spruce, Englemann spruce and subalpine spruce.

Structural lumber is graded according to strength, this has been done scientifically with electronic devices and stamped with information indicating the specific loads it will support. This is called stress-rated lumber.

Stress-rated lumber may be specified at times for specific items where certain strength properties are crucial. For the average house construction, yard lumber—carried by most lumberyards—is usually recommended for residential roof framing. Stress-grade lumber can be obtained on special order if needed.

The preceding grading rules and standards are supplied by West Coast Lumber Inspection Bureau.

MATERIAL SELECTION

When selecting material for any construction project, it is suggested that the following factors be considered.

☐ Consider all grades suitable for the intended use. For practical purposes use the lowest acceptable grade.

☐ Seasoned lumber should be specified for product quality, stability, and nail-holding ability.

☐ Choose the species most suitable for the job.

☐ Specify grade-stamped lumber. The stamp on the lumber purchased assures you that you are buying the material specified.

Storing Lumber

It is a sad fact of life in the construction industry that once most lumber leaves the mill it is not handled properly. All the care taken prior to arriving at the lumberyard and after being received on the job can be nullified by improper storage. Here are some suggestions on the proper handling of lumber:

☐ Try to purchase lumber that is protected by storage sheds and covered on tops, sides, and ends with plastic or another waterproof material.

☐ Lumber received on the job should be stored carefully to prevent shrinking and swelling.

☐ It should be kept off the ground.

☐ Keep it clean.

☐ Pile it neatly with stickers in between layers.

☐ Keep it covered with waterproof material until used on the job site.

Roof-Sheathing Material

When the rafters are in place and the roof has been checked to make sure it is level, plumb, and square, apply sheathing. Roof sheathing performs several important functions. It adds a great amount of strength to a building by helping to tie the building together and by keeping it from wracking. Sheathing also helps protect the structure and supplies a sound surface upon which you can apply the roof covering.

In the past, 1-×-6 random-length boards were the most common material used. These boards have been replaced in popularity today with plywood. Other sheathing materials are available, also. Sheathing boards can be used when certain roof coverings such as wood shingles are to be applied.

Subsequent discussions of sheathing materials in this book are limited to plywood because it is the most practical and most popular product used for sheathing roofs.

Plywood

The use of plywood became widespread after World War II. Plywood had been available prior to this time, but its uses and advantages were not fully realized until its exposure during the war.

Plywood is made up of thin layers of wood called plies (veneers) that are bonded together with special adhesives under great pressure and heat. Alternate plies of wood placed at right angles to each other are also glued and bonded. This arrangement increases the strength of the wood across the grain (an inherent weakness in wood boards). The center plies are called the core, the outer plies are called faces, and the plies in between are called crossbands.

Plywood sheets are composed of three, five, and seven plies. Having the total thickness of the plies running in one direction equal to the thickness of the plies running at right angles helps reduce warping. The standard size of plywood sheets is 4 feet in width and 8 feet in length. These are the common sizes for construction use. Special sizes of 6 to 8 feet wide and 9 to 16 feet in length are available on special order. Common thickness of plywood ranges from 5/32 of an inch up to 1 1/2 inches. Edges may be square cut and butted together, or rabbeted so that they overlap (called shiplap) or interlock with tongue-and-groove edges.

The type of glue used to manufacture plywood controls where plywood can be used in building construction. Interior plywood is made with an adhesive that is water resistant. It will withstand an occasional wetting and should be used in areas where moisture is not present in large amounts. Exterior plywood is made with an adhesive that is waterproof. This material will withstand repeated wetting without affecting its strength or shape. The type of plywood and type of glue used will be identified and listed on the grade stamp.

Most plywood is manufactured from native softwoods, and occasionally some plywood is manufactured from hardwoods. More than 70 species of wood are used in the manufacture of plywood. These species are classified in five groups numbered from one to five (with one being the strongest). Plywood made from fir and southern pine are the most common types of wood used for roof sheathing.

On most grade stamps, two large letters will be found separated by a hyphen. These letters indicate the quality or grade of the faces of the plywood sheets. There are 5 letters used A, B, C, D, and N. Grade (N), the top quality, is free of defects and is used when a natural finish is to be applied. This grade is not intended for roof sheathing.

The letters A through D are used to grade the commonly used panels: Grade (A) being the highest and grade (D) the lowest. All grades can be combined according to the use for which that particular plywood panel is intended. Sheathing is usually classified as CDX. This indicates that the quality of that panel has one side that is grade (C) and the other side is grade (D), with a core of grade (D). The (X) indicates that exterior glue has been used, making it suitable for roof sheathing.

It is best to store plywood flat, raised above the ground, and fully supported under the entire surface across the grain. Thicker panels can be stored with a waterproof covering until used.

RECOMMENDED MATERIALS FOR ROOF CONSTRUCTION

Rafters:
- ☐ Number 2 or better kiln dried (KD) Douglas fir.
- ☐ Number 2 or better seasoned is also an accepted material.
- ☐ Moisture content (MC) approximately 16 percent to 19 percent maximum.

Sheathing:
- ☐ 4-×-8-foot plywood CDX grade.
- ☐ Fir or southern pine, 3/8-inch minimum thickness, over conventional rafter-framed roof.

Fasteners:
- ☐ Rafters should be framed with 16D (penny) common nails.
- ☐ Sheathing should be fastened with 7D (penny) common cement-coated nails.

Note: In high-moisture areas, some building codes require hot-dipped galvanized nails.

PROGRESS CHECK

1. Wood is not a common material for residential construction. T — F.

2. Wood is classified as _____ or _____, depending on which type of tree it comes from.

3. Most construction lumber is softwood. T — F.

4. After boards are cut and trimmed to width they are _____ and _____ to size.

5. Green wood has a moisture content of _____ to _____ percent.

6. Dry wood has a moisture content below _____ percent.

7. MC is the abbreviation for _____ _____ of lumber.

8. Two methods of drying lumber are _____ drying and _____ drying.

9. Air drying is the desired method of drying because it gives better control. T — F.

10. Kiln drying is the faster method of drying lumber. T — F.

11. Lumber is graded in such a way that defects can be concealed from the purchaser. T — F.

12. Lumber is graded in accordance to _____ sets

of rules, known as _____ and _____ grading rules.

13. The _____ _____ _____ is responsible for outlining the grading rules.

14. Most construction lumber comes from the western part of the country. T — F.

15. Always use the lowest acceptable grade for the intended use. T — F.

16. For average house construction, _____ grade lumber is usually used.

17. Is it necessary to store lumber properly on the job site until used? Yes — No.

18. _____ is the most common material used for roof sheathing.

19. _____ _____ are still used as roof sheathing when the covering for the roof is wood shakes.

20. To give plywood its great strength, the grain of the alternate plies are placed at _____ angles to each other.

21. The factor that controls where plywood can be used is the type of adhesive. T — F.

22. Most plywood used in construction is manufactured from _____ _____, with fir and southern pine being the most common.

23. There are _____ letters used to indicate the grade of plywood.

24. The letter (N) indicates that the plywood panel is of the best quality. T — F.

25. The three letters _____ indicate that the plywood sheathing is suitable for roof-sheathing purposes.

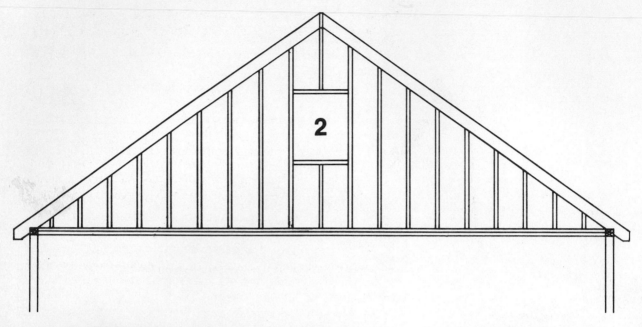

Math for Roof Framing

Objective. Upon completion of this chapter, you should be thoroughly familiar with the operation of an electronic calculator and the following mathematical operations:

- ☐ Addition of whole numbers.
- ☐ Subtraction of whole numbers.
- ☐ Multiplication of whole numbers.
- ☐ Division of whole numbers.
- ☐ Fractions as used in roof framing.
- ☐ Decimals as used in roof framing.
- ☐ Conversion of fractions and decimals.
- ☐ Square root.
- ☐ Use of the electronic calculator.

Take the time to fully review and practice all the types of problems shown in this chapter. The time spent now will make the learning of roof framing much easier. The math reviewed in this chapter is basic to the understanding of roof framing.

WHOLE NUMBERS

Whole numbers are integers. Integers are any of the natural numbers, negatives of these numbers, or zero.

Addition

Addition is the process by which we find the sum of two or more numbers. The following example shows the correct arrangement of an addition problem.

$$
\begin{array}{r}
32568 \\
4809 \\
+\ 102 \\
\hline
37479
\end{array}
$$

Write the numbers one under the other. Align the digits. Add each digit in a vertical direction, placing the sum under the column added. This process is repeated until all the columns have been added giving a total sum.

Subtraction

Subtraction is the process of taking away one number from another.

$$
\begin{array}{r}
57891 \\
-\ 3241 \\
\hline
54650
\end{array}
$$

To subtract whole numbers, write the larger of the two numbers (minuend) on top, start with the right-hand column and subtract the smaller numbers from the larger numbers. Keep the digits in the answer aligned under the columns in the problem.

Multiplication

Multiplication is a short method of repeated addition.

$$
\begin{array}{r}
426 \\
\times\ 5 \\
\hline
2130
\end{array}
$$

It is much easier to multiply numbers if the larger number (multiplicand) is placed on top and the smaller number (multiplier) is placed underneath, and the extreme right hand columns are kept aligned.

Division

Division is the process of finding out how many times one number is contained in another. Shown on next page is the arrangement and names of a division problem.

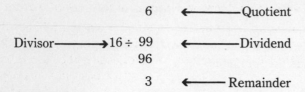

$$
\begin{array}{r}
6 \quad\longleftarrow\text{Quotient} \\
\text{Divisor}\longrightarrow 16 \div 99 \quad\longleftarrow\text{Dividend} \\
96 \\
3 \quad\longleftarrow\text{Remainder}
\end{array}
$$

The number to be divided (dividend) is placed inside the division frame; the number by which the dividend is to be divided (divisor) is placed outside and to the left of the frame. Determine how many times the divisor will go into the dividend and place that number above the division frame to give the answer (quotient). When the divisor does not divide into the dividend evenly, the number left over is called the remainder.

To check a division problem, multiply the quotient by the divisor. The product of these numbers will be the dividend. If there is a remainder add it to the product for a total.

FRACTIONS

A common fraction is a method of dividing a quantity into a number of equal parts. The common fraction is written with two numbers, one placed over the other. The lower number, called the denominator, tells how many parts the quantity, unit, or number has been divided into. The upper number, called the numerator, indicates the number of parts of the quantity, unit, or number being used. Here is a common fraction, and an explanation of the parts of the fractions.

$$
\begin{array}{l}
\text{Numerator}\longrightarrow \qquad \dfrac{1}{3} \\
\text{Denominator}\longrightarrow
\end{array}
$$

The denominator (3) indicates that number or quantity has been divided into three parts. The numerator (1) indicates that only one part of three is being used.

Proper Fractions

Any quantity less than 1 is called a proper fraction; 1/2, 1/8, 1/4, 9/16, 3/32 are some examples of proper fractions. The numerator in a proper fraction is always smaller than the denominator.

Improper Fractions

Any fraction with a quantity greater than 1 is called an improper fraction; 6/4, 5/3, 8/5 are some examples of improper fractions. The numerator in an improper fraction is always larger than the denominator.

34

Reducing Common Fractions to Lowest Terms

The handling of fractions is simplified when you reduce all fractions to their lowest terms. A fraction is said to be in its lowest terms when the denominator or numerator cannot be divided evenly by any other numbers. One of the characteristics of a common fraction is that the numerator and denominator can be divided by the same number without changing its value.

8/16 reduced to lowest terms = 1/2

Both numerator and denominator were divided by 8. This reduced the fraction to 1/2 (its lowest terms).

6/16 reduced = 3/8

Once again the numerator 6 and the denominator 16 were divided by the same number (2), reducing the fraction to its lowest terms. Let me point out that any number can be used, not necessarily just the number 2.

Reducing Improper Fractions

To reduce improper fractions to their lowest terms, divide the numerator by the denominator. Any resulting fraction in the answer then is reduced to its lowest terms.

6/4 reduced to lowest terms = 1 1/2

6/4 reduced to the mixed number 1 2/4 is accomplished by dividing the numerator 6 by the denominator 4. The result is the mixed number 1 2/4. Reducing 2/4 to lowest terms is accomplished by dividing both numerator and denominator by 2, resulting in an answer of 1 1/2.

5/3 reduced to lowest terms = 1 2/3

Dividing the denominator 5 by the numerator 3 gives a result of 1 2/3. The fraction 2/3 is in its lowest terms, resulting in the answer of 1 2/3.

The Common Denominator

When adding or subtracting fractions, a common denominator must first be determined. In carpentry, this is not difficult to do because most of the fractions a carpenter uses deal with the standard scale of inches and fractions of an inch. These fraction are 16ths, 8ths, 4ths, 1/2, and, on rare occasions, 32nds of an inch. Most times a common denominator can be determined by inspection. If this proves difficult, try using the example that follows for an easy method of finding a common denominator.

$$1/3 + 5/16 + 3/4 + 5/8 = 48$$

1. Arrange the denominators in a horizontal row separated by dashes:

$$3 - 16 - 4 - 8$$

2. Divide these denominators by a prime number that goes into at least two of the denominators without leaving a remainder. In this problem, 4 fills that requirement.

$$\frac{4)\ 3\ -\ 16\ -\ 4\ -\ 8}{\quad 4 \quad\ 1 \quad\ 2}$$

3. Bring down any numbers that do not divide evenly. In this problem the number is 3.

$$\frac{4)\ 3\ -\ 16\ -\ 4\ -\ 8}{\quad 3 \quad\ 4 \quad\ 1 \quad\ 2}$$

4. When necessary, repeat this process until there are no two numbers remaining that can be divided without leaving a remainder.

$$\frac{2)\ 3 \quad\ 4 \quad\ 1 \quad\ 2}{\quad\ 3 \quad\ 2 \quad\ 1 \quad\ 1}$$

5. Multiply the vertical column ($2 \times 4 = 8$) by the horizontal row ($3 \times 2 \times 1 \times 1 = 6$). The product of this step is the lowest common denominator (48).

This process for finding the common denominator of a group of fractions can be used regardless of the number of fractions, or value of the denominators in the problem.

Addition of Fractions

To find the common denominators for all fractions, use the inspection method or the one described above to accomplish this.

Add all numerators.

Write the sum over the common denominator.

If necessary reduce the answer to its lowest terms.

$$\frac{1}{4} + \frac{5}{8} + \frac{7}{16} + \frac{3}{4} = 2\frac{1}{16}$$

In this example, find the common denominator by inspection. The inspection method is best for this problem because 16 will divide evenly into all the numerators, and this can be determined just by observing the example.

With 16 as the common denominator both the numerator and denominator are multiplied by the same number, the product of which will be 16.

$$\frac{1}{4} \times \frac{4}{4} = \frac{4}{16}$$

$$\frac{5}{8} \times \frac{2}{2} = \frac{10}{16}$$

$$\frac{7}{16} \times \frac{1}{1} = \frac{7}{16}$$

$$\frac{3}{4} \times \frac{4}{4} = \frac{12}{16}$$

Add the numerators and place over the common denominator 16.

$$4 + 10 + 7 + 12 = \frac{33}{16}$$

Reduce the answer to lowest terms if necessary. In this example, dividing 33 by 16 results in an answer of 2 with a remainder of 1/16.

$$33 \div 16 = 2\ 1/16$$

Subtraction of Proper Fractions

The steps in the subtraction of proper fractions are as follows:
    a. Change all fractions to the lowest common denominator.
    b. Subtract the numerators.
    c. Place the difference over the lowest common denominator.
    d. Express the resulting fraction in lowest terms.

$$\frac{12}{16} - \frac{5}{8} = \frac{1}{8}$$

a. Determine the lowest common denominator by inspection; it is 16.

b. Multiply the numerators and denominators of the fractions by the same number that will give a produce of 16.

$$\frac{12}{16} \times \frac{1}{1} = \frac{12}{16}$$

$$\frac{5}{8} \times \frac{2}{2} = \frac{10}{16}$$

c. Subtract the numerators and place over the common denominator (16) for a result of 2/16.

$$\frac{12}{16} - \frac{10}{16} = \frac{2}{16}$$

d. Reduce 2/16 to lowest terms by dividing both numerator and denominator by 2.

$$2/16 \div 2/2 = 1/8$$

Multiplication of Fractions

Steps in the multiplication of fractions:

a. Multiply the numerators.

b. Multiply the denominators.

c. Place the product of the numerators over the product of the denominators.

d. Reduce the fraction to lowest terms if necessary. Note that no common denominator is needed when multiplying fractions.

$$\frac{3}{8} \times \frac{3}{4} = \frac{9}{32}$$

a. Multiply the numerators 3 × 3 = 12.

b. Multiply the denominators 8 × 4 = 32.

c. Place the product of the numerators (9) over the product of the denominators (32) for an answer of 9/32.

d. Reduce the fraction to lowest terms, if necessary. In this example no reduction is required.

Division of Fractions

Steps in the division of fractions:

a. Invert the divisor which is the dividing fraction. This is the second fraction in the problem (that fraction that follows the division sign).

b. Change the division sign to a multiplication sign and multiply, following the procedure described under multiplication of fractions.

$$3/8 \div 3/4 = 1/2$$

a. Invert the divisor and change signs.
   $3/8 \div 3/4$ is inverted to $3/8 \times 4/3$.
b. Multiply the numerators $3 \times 4 = 12$.
c. Multiply the denominators $8 \times 3 = 24$.
d. Arrange the numerator over the denominator 12/24.
e. Reduce the result to lowest terms by dividing the numerator and denominator by 12.

$$\frac{12}{24} \div \frac{12}{12} = \frac{1}{2}$$

DECIMAL FRACTIONS

Fractions stated as a decimal are also called decimal fractions. Any fraction, such as 1/2, 1/4, 2/3, 3/16, can be expressed as a decimal. This is accomplished by dividing the denominator into the numerator. The result will be a group of numbers preceded by a decimal point.

The following are examples of fractions converted and expressed as decimals:

$1/2 = 1 \div 2$ or .5 spoken as point five or 5 tenths.
$1/4 = 1 \div 4$ or .25 spoken as point 25 or 25 hundredths.
$3/8 = 3 \div 8$ or .375 spoken as point 375 or 375 thousandths.
$2/3 = 2 \div 3$ or .66 spoken as point 66 or 66 hundredths.

Note that when a number such as .66 repeats itself infinitively it is known as a repeating decimal. A decimal fraction can be carried out to any number of places, depending on the degree of accuracy desired. In carpentry, the decimal fraction is usually carried out to two decimal places.

Mixed Numbers Containing
Common Fractions Expressed As Decimals

Any mixed number containing a common fraction can be expressed as a whole number with the fraction in decimal form. The mixed number 56 1/4, when converted to decimals, becomes 56.25; it is composed of the whole number 56 (to the left of the decimal point) and a decimal fraction .25 (with the numerals 25 immediately following the decimal point to the right). The following are examples of mixed numbers converted to decimals:

$$4 \ 1/2 = 4.5$$
$$3 \ 3/4 = 3.75$$
$$5 \ 1/3 = 5.33$$

The decimals .5, .75, and .33 were found by following the procedure described in the preceding example (fractions converted and stated as a decimal), and then placing it after the whole number.

The Value of Zeros and the Decimal Point

A single zero placed before the decimal point does not change the value of the problem. This placing of the zero is done to eliminate confusion and to emphasize the fact that the number is less than one (therefore making it a decimal fraction).

$$0.45$$
$$0.367$$
$$0.079$$

The placing of zeros after the last numeral in a decimal does not change the significant value of that number.

$$0.9870$$
$$0.500$$
$$0.0250$$

The use of zeros does not change the value of the decimals in the above examples. They are still spoke as 987 thousandths, 5 tenths, and 25 thousandths.

At times it will be necessary to multiply decimals that will result in a product that does not have enough numerals to place the decimal point in its proper place. When this occurs, it is necessary to add zeros to the left of the first significant numeral. Multiply and count off the proper number of decimal places:

$$\begin{array}{r} .025 \\ \times \ .05 \\ \hline 125 \end{array}$$

When placing the decimal point, it is necessary to count the number of decimal places in the problem, and then in the product. Starting from the extreme digit on the right, count off the required number of decimal places and place the decimal point. In this problem, there are not enough digits in the answer to place the decimal point in its proper place. It then becomes necessary to add the required zeros to the left of the first digit in the answer. Counting off it is found that this problem requires five decimal places. With only three places in the answer, it then becomes necessary to add two zeros in front of the first numeral—creating an answer of .00125.

Rounding Off

The accuracy or tolerance to which a craftsman works is controlled by the number of decimal places a mathematical calculation is being carried out to. Those who work in the machine trades are required to work to tolerances of thousandths (three places) and at times hundred thousandths (four places) of an inch. In carpentry, a tolerance of one hundredth of an inch (two places) is an accepted standard. For example, 12.41 is carried out to two places as opposed to 12.4132, a number carried out to four places. The process of obtaining the required number of decimal places is called rounding off.

Here is an example of rounding off to two places:

$$0.44769$$

a. Locate the second digit (4) to the right of the decimal point.

b. Observe the third digit to the right of the decimal point. If it is less than the number five, drop that digit and all the digits to the right of it that follow. If the third digit is greater than the number five, increase the numeral to the left by one number and drop the third digit and all those that follow it to the right.

c. In this problem, it is found that the third digit is the number 7. Following the preceding rule, the numeral in the second place is increased by one number and the 7 and all the numerals to the right are dropped. This gives an answer of 0.45 (which is 0.44769 rounded off to two places).

Practice rounding off the following numbers:

0.332064	0.005218	1.256344
0.569310	0.193398	2.74961

You will find that in roof framing rounding off decimals is quite common and that rounding off to two places is an accepted standard in the carpentry trade. Keep in mind that in rough carpentry a margin of error of 1/8th of an inch is an accepted amount of error.

Converting Fractions to Decimal Fractions

In solving roof framing problems, it is necessary to work with both fractions and decimals. All measurements that a carpenter works with are given in feet, inches, and fractions of an inch. These units are shown on a standard rule or tape. In the calculation of the various parts to be used, some information is given in the decimal equivalents. For this reason, the carpenter must become adept at converting the given information into a usable form.

The steps in changing fractions of an inch to decimal fractions of an inch are as follows.

a. Divide the numerator by the denominator.

b. The answer will be in decimal form. In most cases it will be more practical to round off the answer to two places. Convert the following decimal and round off to two places.

$$1/16 = 1/16 \div 16 = .06$$

$$
\begin{array}{r}
.0625 \\
16\,\overline{)1.0000} \\
\underline{96} \\
40 \\
\underline{36} \\
40 \\
\underline{36} \\
4
\end{array}
$$

Round off to two places. The third digit (2) is less than 5. Following the rule for rounding off, the numeral 2 and all following numerals to the right of it are dropped, giving an answer of .06.

Convert the following and round off to two places:

1/32
1/8
1/4
5/8
9/16

Converting Inches to Decimals of a Foot

To convert any inch increment between 1 and 12 inches (1 foot) to a decimal of a foot, arrange the problem with the inch digit as the numerator and the digit 12 as the denominator.

1/12, 2/12, 3/12, 5/12

To convert 5/12 to decimals of a foot:

$$
\begin{array}{r}
.4166 \\
12\overline{)5.0000} \\
4\,8 \\
\hline
20 \\
12 \\
\hline
80 \\
72 \\
\hline
80 \\
80 \\
\hline
8
\end{array}
$$

Following the preceding example, construct a table converting each 1-inch increment on a rule from a fraction to a decimal equivalent; begin with 1/12 and end with 12/12.

Converting Fractions of an Inch to Decimal Fractions of an Inch

To express a decimal fraction as a common fraction, multiply the decimal by whichever fraction you want to express it in (such as 32nds, 16ths, 8ths). Here is an example of converting a decimal fraction to a common fraction:

0.25 converted to 16ths of an inch

$$ .25 = \frac{.25}{1} \times \frac{16}{16} = \frac{.25 \times 16}{1 \times 16} = \frac{4}{16} = \frac{1}{4} $$

Convert the following decimal fractions to common fractions:

0.33
0.42
0.50
0.56
0.75

Short Method of Converting Decimals of an Inch to Fractions of an Inch

In practical use, it is only necessary to multiply the decimal by the value of the fraction it is to be changed to. In other words, to convert a decimal fraction to 1/8ths of an inch multiply the decimal by 8 and place the result over 8, creating a fraction. To convert a decimal to 16ths, just multiply the decimal by 16 and place the answer over 16.

Examples in converting decimals to common fractions of an inch using the short method.

0.5 × 64 = 32/64 reduced to 1/2
0.33 × 8 = 2.64/8 rounded off to 3/8

Change the following decimals to the given fractional values.

0.25 to 4ths
0.75 to 4ths
0.125 to 8ths
0.42 to 32nds
0.53 to 64ths

SQUARE ROOT

Even though the computation of square root is explained in this chapter's section on the use of the calculator, it would be remiss of me not to explain how to determine the square root of a number arithmetically. At first this method of computing square root might seem complicated and confusing, but with a little practice it does become quite easy, although, I must admit, tedious at times.

Determining Square Root

$$\sqrt{1\ 80\ .53\ 81\ 00}$$

The number from which the square root is to be extracted is placed inside the radical sign (which is the symbol for square root) and divided into pairs, starting from the decimal and working left and right. The extreme left-hand pair may consist of only one numeral, but it is still considered and handled as a pair. As in division, if it is desired to carry the remainder out to a greater number of places, pairs of zeros are added after the last numeral pair on the right and the square root process is continued. Notice that the number of digits in the answer is determined by the number of pairs in the dividend (one digit in the answer for each pair in the dividend).

The Parts of a Square Root Problem

```
                    1 3 .4 1 6  ←————————Root (quotient)
Radical Sign————→  √1 80.00 00 00  ←————————Dividend
                   1| 1
                    ‾‾‾‾‾
Trial Divisor————→ 20)
                   +3  0 80  ←————————New Dividend
                   ‾‾‾‾‾‾‾
New Divisor————→ 23
```

44

To find the square root of 561:

a. Place the number in the radical sign and separate into pairs.

$$\sqrt{5\ 61.}$$

There are no significant numbers after the decimal point, indicating that the answer should consist of two digits (the number of pairs in the radical sign). Unless there is a remainder, no zeros will have to be added to the dividend after the decimal point.

b. Find a number that, when multiplied by itself (squared), will not be larger than the first pair of numerals in the dividend (in this problem it is 2). The 2 is placed in the quotient (root) above the first pair in the dividend. It is then squared and placed under the first pair of numbers (5) in the dividend.

c. The second pair of numbers (61) is brought down to give a new complete dividend of 161.

d. Multiply the 2 in the quotient by 20 to find a trial divisor of 40. Divide this trial divisor of 40 into the new dividend of 161, which gives a numeral 3 to be placed in the quotient along side the numeral 2. Add the numeral 3 to the trial divisor of 40 for a new divisor of 43. Multiply the divisor of 43 by the numeral 3 in the quotient, which will give 129 to be placed under the dividend of 161 in the problem, and subtract for a new dividend of 32. Bring down the two zeros to the right of the decimal point for a complete dividend of 3200.

e. Multiply the quotient of 23 by 20 to get a new trial divisor of 460, which will go into 3200 the new dividend 6 times. The numeral 6 is placed in the quotient and also added to the new trial divisor to get a divisor of 466, which is multiplied by the numeral 6 in the quotient. The product (2796) is then placed under the 3200 and the subtracted to give a dividend of 404, bring down the next pair of zeros to get a new complete dividend of 40400.

f. Multiply the quotient of 236 by 20 to get a new trial divisor of 4720, which will go into 40400 the new dividend 8 times. The numeral 8 is placed in the quotient and also added to the new trial divisor to get a new divisor of 4728, which is multiplied by the numeral 8 in the quotient. The product (37824) is then placed under 40400 and subtracted to give a new dividend of 2576, bring down the next pair of zeros to get a new complete dividend of 257600.

g. Multiply the quotient of 2368 by 20 to get a new trial divisor of 47360, which will go into 257600 the new dividend 5 times. The numeral 5 is placed in the quotient and added to the trial divisor to get a new divisor of 47365, which is multiplied by the numeral 5 in the quotient. The product (236825) is placed under 257600 and subtracted to give a new complete dividend of 38775.

This process can be carried on to create any desired number of decimal places, depending on the accuracy to which you must work. For our purpose, three decimal places is sufficient. Therefore, I will end the square root process at this point. To prove the work just completed, square the answer 23.685 (multiplied by itself) and the result will be 560.979 (rounded off to 561).

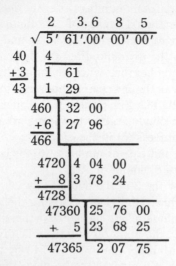

THE ELECTRONIC CALCULATOR

Man always has had some form of calculator. Fingers provide the most primative form of calculator. Common calculating devices are the abacus, the slide rule, and the electronic calculator.

The calculator should be considered a tool. The same way a power saw helps do the job quicker than a hand saw so does the calculator do mathematical problems quicker than the long-hand method. It is important, however, that you understand the theory and methods of the problem you are trying to solve.

While it is not the intent or purpose of this book to teach the full use of calculators, I will explain those operations of the calculator needed to solve roof framing problems. Before you begin calculations, review the instruction book that comes with your calculator.

Objective

Upon completion of this section, you should be familiar with the basic fundamentals of a calculator, how to solve basic arithmetic problems, and especially those problems used in roof framing. To explain the basic operation of a calculator I have chosen a Radio Shack Electronic Calculator Model EC-205. It is a relatively inexpensive, battery-operated unit (Fig. 2-1) with the following functions.

1. Addition
2. Subtraction
3. Multiplication
4. Division
5. Square Root
6. Percentage
7. Memory

Most calculators operate on the same principle, but you will find in general that as units become more expensive they are intended to solve more complicated mathematical and scientific problems. For basic roof framing, and most problems that have to be solved by a carpenter, this type of calculator will serve you well.

Operation of the Calculator

The model EC-205 calculator is battery operated. Some calculators are powered by battery or house current. Others are solar or light operated. The rectangular window at the top is the display. The keyboard consists of 24 keys. This is typical for a calculator in this price range. The keys are labeled in accordance to the function they perform.

ON/C CE Turns the calculator on. Press once to clear wrong entry. Press twice or more to clear the calculator.

Off Turns the calculator off.

Number Keys Enter into the display any number pressed in the exact sequence pressed.

. Inserts a decimal point in the problem at the place it is entered.

+, −, ×, and ÷ Performs the operation they indicate.

% Used to calculate percentage. A very limited use in roof framing.

$\sqrt{\times}$ Determines the square root of any number showing on the display.

M+, M−, MR, MC Stores in memory any numbers entered into it. Memory performs a subtotal function. Any information can be recalled (MR) when needed. Information can be added or subtracted (M+, M−) as required by the problem. The memory can be cleared by the (MC) key without affecting the primary problem. Memory will be used by the carpenter to determine square root.

= Total key commands the calculator to perform the desired function and display the result.

Addition

Use the calculator to solve the following problems. Check the

answers by doing the problem long hand. All operations on the keyboard are listed under functions. All numbers entered on the keyboard will be shown under the display heading.

Add the following whole numbers:

$$253 + 695 + 21 + 3525 = 4494$$

FUNCTION	DISPLAY
Press (on)	(0)
Enter 253	(253)
Press +	(253)
Enter 695	(695)
Press +	(948)
Enter 21	(21)
Press +	(969)
Enter 3525	(3525)
Press =	(4494) Answer

Add the following decimal numbers:

$$57.25 + 951.3 + 6.5732 = 1015.1232$$

FUNCTION	DISPLAY
Press (on)	(0)
Enter (57.25)	(57.25)
Press +	(57.25)
Enter 951.3	(951.3)
Press +	(1008.55)
Enter (6.5732)	(6.5732)
Press =	(1015.1232)

Observe that the calculator automatically aligns the decimal point. To solve any problem, it is necessary to set the problem up in the proper sequence of steps. Enter the numbers and functions in the order that they are written and give the calculator time (a matter of fractions of a second) to compute the answer.

Subtraction

Subtract the following decimal numbers:

$$4561 - 352.25 = 4208.75$$

FUNCTION	DISPLAY
Press (on)	(0)
Enter 4561	(4561)

Fig. 2-1. Left: A basic electronic calculator is an ideal instrument to learn the fundamentals of operating a calculator. Courtesy of Radio Shack, Division of Tandy Corp.

FUNCTION	DISPLAY
Press –	(4561)
Enter 352.25	(352.25)
Press =	(4208.75) Answer

Once again you will observe that the calculator automatically aligns the decimal points and does the necessary computation to obtain the correct answer.

Multiplication

Multiply the following whole numbers:

$$6745 \times 193 = 1,301,785$$

FUNCTION	DISPLAY
Press (on)	(0)
Enter 6745	(6745)
Press ×	(6745)
Enter 193	(193)
Press =	(1,301,785) Answer

Your new-found tool has done it again. By entering the given information, you will receive the correct answer in a fraction of a second. Bear in mind that the accuracy is controlled by the information it receives.

Division

Divide the following numbers:

$$Divide\ 6397.48 \div 15 = 426.4986$$

FUNCTION	DISPLAY
Press (on)	(0)
Enter 6397.48	(6397.48)
Press ÷	(6397.48)
Enter 15	(15)
Press =	(426.4986)

With subtraction, always enter the dividend first, then the function key, followed by the divisor. Press the total key (=) to display the result.

At this time, I suggest that you complete the following problems. This will give you the practice needed to feel comfortable with the four basic (+, −, ×, ÷) functions of the calculator.

Addition

5,921 + 33,465 + 25 + 167 =

25 + 14.55 + 7981.32 + 50.05 =

23.47 + 90.51 + 85.05 + 55 =

53 + 69 + 75 + 150 + 95 =

357,891 + 253,355 + 15,034 =

Subtraction

65,480 − 3,552 =

751 − 25 =

244,873 − 145.75 =

275.45 − 145.75 =

4,598.3 − 1,690 =

Multiplication

55 × 75 =

579 × 2,541 =

15.49 × 20 =

16.35 × 8.75 =

1,595 × 20 =

Division

250 ÷ 5 =

4,353 ÷ 97 =

15 ÷ 150 =

9,565.374 ÷ 37.08 =

5,275 ÷ 25 =

Do you feel at ease with the calculator? If you do move on to the next part, if not practice some more, make up your own problems if you have to. Practice until you are sure of all the operations to this point.

Memory

Memory is useful when multifunction operations are required, as in the solution of formulas. If a calculator does not have memory, (subtotal) it is necessary to record each subtotal operation on paper and then reenter these results to obtain the answer. With a calculator that has memory, the information is stored in the calculator and recalled when needed.

When a series of operations involve addition, subtraction, multiplication, and division, they must be performed in a definite order. Here is the arithmetical order of problem solving:

1. Perform any multiplication.

2. Perform any division.

3. Any addition or subtraction is performed in the same order they occur in the problem.

4. Note if any operations are contained between parentheses they must be removed first using the same order listed in the steps above.

Use of Parentheses

In practical application, parentheses are used to define the order of operation and reduce the chances of error.

$$\text{Solve } 6 + 7 \times 3 = 27$$

The above example should have been written with the use of parentheses $6 + (7 \times 3) = 27$. Notice that the parentheses clearly defined the fact that the multiplication had to be performed in accordance to the arithmetical order.

Solve the problem $9 + (4 \times 5) = 29$. Following the order of operation, 4×5 is multiplied first. The resultant product is then added to 9 for a total of 29.

FUNCTION	DISPLAY
Press (on)	(0)
Enter 4	(4)
Press ×	(4)
Enter 5	(5)
Press m+	(m 20)
Enter 9	(9)
Press mr	(m 29) Answer

Notice the arithmetical order (multiplication first then addition). Solve the problem $(8 \times 6) - (4 + 5) = 39$

FUNCTION	DISPLAY
Press (on)	(0)
Enter 8	(8)
Press ×	(8)
Enter 6	(6)
Press m+	(m 48)
Enter 4	(4)
Press +	(4)
Enter 5	(5)
Press m−	(m 9)
Press mr	(m 39) Answer

Once again, notice that the arithmetical order has been followed (multiplication, addition, then subtraction).

Using Memory

Calculate $(159 \times 33.7) - 34.25$ plus $(75 \times 5) - 93 = 5{,}606.05$

FUNCTION	DISPLAY
Press (on)	(0)
Press mc	(0)
Enter 159	(159)
Press $\times$	(159)
Enter 33.7	(33.7)
Press m +	(m 5,358.3)
Enter 34.25	(m 34.25)
Press m –	(m 34.2)
Enter 75	(m 75)
Press $\times$	(m 75)
Enter 5	(m 5)
Press m +	(m 375)
Enter 93	(m 93)
Press m –	(m 93)
Press mr	(m 5,606.05) Answer

Calculate $(255.71 \times 30) - 502$ plus $(351 \times 15.85) - 101 = 12{,}631.65$

FUNCTION	DISPLAY
Press (on)	(0)
Press mc	(0)
Enter 255.71	(255.71)
Press $\times$	(255.71)
Enter 30	(30)
Press m +	(m 7,671.3)
Enter 502	(m 502)
Press m –	(m 7,169.3)
Enter 351	(m 351)
Press $\times$	(m 351)
Enter 15.85	(m 15.85)
Press m +	(m 5,563.35)
Enter 101	(m 101)
Press mr	(m 12,631.65) Answer

Practice problems using memory

$15 + (6 \times 7) + (8 \times 9) - 25 =$
$(9 \times 8) + 75 - 39 + 21 =$
$(7 + 5) \times 8 - (4 - 2) =$

$(12 \times 3) \div 2 - (4 + 8) =$
$(5 \times 9 \div 2) + 37 - (3 \times 2) =$

SQUARE ROOT ON A CALCULATOR

A carpenter will be required to solve a number of square root problems. With an electronic calculator it becomes a matter of entering the number to be squared into the calculator and pressing the square root key.

$$\sqrt{175} = 13.22876$$

FUNCTION	DISPLAY
Press (on)	(0)
Press mc	(0)
Enter 175	(175)
Press $\sqrt{x}$	(13.22876) Answer

Find the square root of the following numbers:

25 =
144 =
345.17 =
36784.532 =
6789 =

Square Root Using Memory

$$\sqrt{65.25^2 + 35^2} = 74.044$$

FUNCTION	DISPLAY
Press (on)	(0)
Press mc	(0)
Enter 65.25	(65.25)
Press ×	(65.25)
Enter 65.25	(65.25)
Press m+	(m 4257.5625)
Enter 35	(m 35)
Press ×	(m 35)
Enter 35	(m 35)
Press m+	(m 1225)
Press mr	(m 5482.5625)
Press $\sqrt{x}$	(m 74.044) Answer

Find the square root of the following

$$\sqrt{6.5^2 + 3^2} =$$
$$\sqrt{4.33^2 + 6^2} =$$
$$\sqrt{12^2 + 12^2} =$$
$$\sqrt{15.55^2 + 25^2} =$$
$$\sqrt{125^2 + 103^2} =$$

Percentage

There is no need for the percentage function in the calculation of roof framing members. You may find it useful in estimating the amount of material needed to construct the roof. The experienced estimator always adds a percentage of the required material to the estimate. This is to cover any loss due to waste.

To compute percentage, the following procedure is followed. Assuming that the estimated number of 2- × -6 roof rafters comes to 1565 lineal feet, with an estimated waste factor of 10 percent, it will be necessary to order in 1721.5 (or 1722, after rounding off) lineal feet.

FUNCTION	DISPLAY
Press (on)	(0)
Enter 1565	(1565)
Press +	(1565)
Enter 10	(10)
Press %	(1712.5) Answer

The calculator automatically does the necessary multiplication and addition to compute the percentage and displayed the correct answer of 1,721.5 lineal feet.

To deduct a percentage amount the subtract (–) key is used in place of the add (+) key. Here is an example of subtracting a percentage. With this example, the same figures in the addition example are used.

FUNCTION	DISPLAY
Press (on)	(0)
Enter 1565	(1565)
Press –	(1565)
Enter 10	(10)
Press %	(1408.5) Answer

Remember that the calculator is a tool that is only as accurate as the operator. If fundamental arithmetic is a weak point for you, take the time to review your answers. Used properly, the electronic calculator can be a great asset in your work. It will increase the speed and accuracy of all calculations on the job, but it will never replace the fundamental principles of math needed to understand the problem being solved.

PROGRESS CHECK

1. 46,957 + 1,321 + 652 + 2,345 =
2. 59,873 − 24,971 =
3. 234 × 21 =
4. 69,871 ÷ 15 =
5. 45.31 + 88.62 + 1.51 + 25.0 =
6. 33.09 − 15.42 =
7. 21.33 − 15.92 =
8. 4972.28 ÷ 7 =
9. 12/16 reduced to lowest terms =
10. 4/12 reduced to lowest terms =
11. 8/5 reduced to lowest terms =
12. Find the common denominator and add, 2/3 + 1/4 + 3/16 + 7/8 =
13. Find the common denominator and subtract, 5/8 − 7/16 =
14. Multiply, 5/16 × 3/4 =
15. Divide, 3/16 ÷ 3/4 =
16. The square root of, 125 =
17. The square root of, 144 =
18. The square root of, 169 =

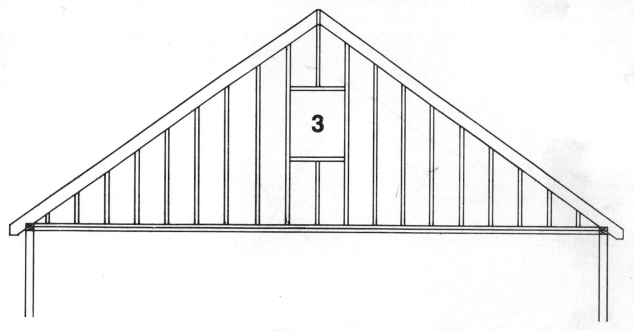

The Steel Framing Square

Objective. Upon completion of this chapter, you should be fully familiar with the steel framing square and how it is applied to the geometry of roof framing. The following information is covered in this chapter:

- [ ] A description of the square and the tables and scales on it.
- [ ] How to use the tables and scales that apply to roof framing.
- [ ] The principles of the right triangle.
- [ ] Similar right triangles.
- [ ] The Pythagorean theorem.
- [ ] Laying out angles with the square.
- [ ] Using the 1/12th scale.

The origin of the framing square is lost in time. No one knows for sure when it was first used by carpenters in the building trades. The square, in some form, was used by our ancient civilizations in the Middle East, and it has been in constant use since. Early squares were made from materials such as wood, bronze, and iron. Early in the nineteenth century, with the coming of the machine age, the steel square became most popular. Today we have squares made of steel, stainless steel, and aluminium.

In early times, the square was first used to check the accuracy of the carpenter's work and help him lay out square and true cuts.

With the passing of time, it was realized that the square had many more uses. With the application of the geometry of the right triangle and the use of the principles of ratio and proportion, many mathematical problems can be solved. In this book, I limit the application of the square to the solving of basic roof-framing problems.

DESCRIPTION OF THE FRAMING SQUARE

The framing square has two parts (Fig. 3-1): a short leg that is 16 inches long by 1 1/2 inches in width, called the tongue, and a longer leg that is 24 inches long by 2 inches in width, called the body (also referred to as the blade). The point where the two legs meet is called the heel. This is the most common square in use today. There are other sizes but they are not popular and used very seldom. For that reason they will not be described in this text.

Holding the square in your right hand with your body extending to the left, you will be looking at the front of the square. The name of the manufacturer will usually be stamped on the heel of the square. Both the inside and outside edges are graduated in inches and fractions of an inch. The front of the body of the square has the rafter tables stamped on it.

Holding the square in your left hand with your body extending to the right, you will be looking at the back of the square. As with

Fig. 3-1. A standard framing square, scales, tables, and graduations on the face and back of the square. Courtesy of Stanley Tools.

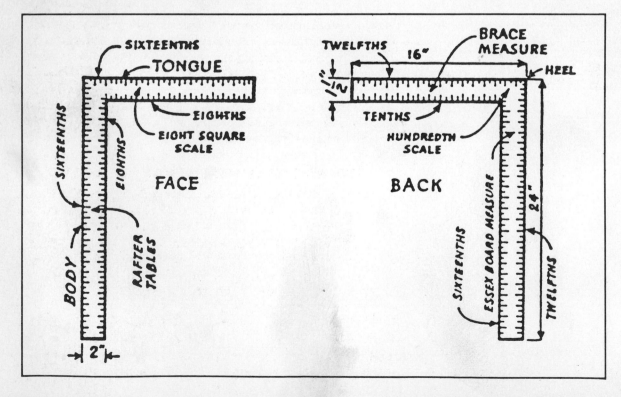

the front of the square, the inside and outside edges are graduated in inches and fractions of an inch. Located on the heel is the 100th scale. The octagon table, the essex board measure, and brace tables are not discussed in this text because they are not used in roof framing.

GRADUATIONS ON THE SQUARE

As shown in Fig. 3-1, the graduations on the edges of the square are divided into the following increments:

> Face of body: outer edge—inches and 1/16ths.
> Face of body: inner edge—inches and 1/8ths.
>
> Back of body: outer edge—inches and 1/12ths.
> Face of body: inner edge—inches and 1/16ths.
>
> Face of tongue: outer edge—inches and 1/16ths.
> Face of tongue: inner edge—inches and 1/8ths.
>
> Back of tongue: outer edge—inches and 1/12ths.
> Back of tongue: inner edge—inches and 1/10ths.

Stamped on the back of the square at the heel is the 100th scale. The upper scale is 1 inch divided into 100 parts; each fine line is actually 1/100th of an inch. Just below it is 1 inch divided into 16ths.

Using a pair of fine-pointed dividers, you can convert from decimal parts of an inch to fractions of an inch. This is accomplished by counting off and setting the dividers to the decimal parts on the upper scale. Without changing the setting of the dividers, move them down and read the fractional equivalent on the 1/16th scale right below it.

Using the square, the tables, and the scales described here and with the application of basic arithmetic, most any problem in roof framing can be solved. See Fig. 3-2.

RAFTER TABLES AND THE RIGHT TRIANGLE

On the face of the body of the square are six rafter tables that are used to calculate the lengths of the rafters, and how to obtain the angle of the various cuts. The use of each line is indicated on the left end of the body (Fig. 3-3). To eliminate any misunderstanding, I will describe the use of line one in detail. This is the only table needed to frame the basic roof. The other five lines are used to frame more complicated roofs such as the hip roof and intersecting hip and valley roofs.

Line one gives the per-foot run of a common rafter; this measurement is in inches. It is the length of the hypotenuse of a right triangle, where the base of the triangle is one foot (12 inches) of run. The altitude (unit rise) is a variable number; this unit rise

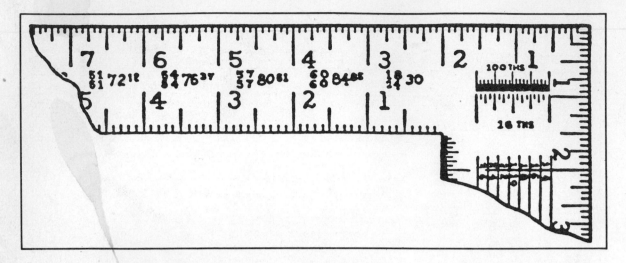

is indicated by the inch graduation marks on the edge of the square. For example, using the unit rise of 8, you will find the number 14.42 on line one directly under the 8-inch graduation on the square. The figure 14.42 is the hypotenuse of the right triangle formed by the unit, rise of 8 and the unit run of 12. See Fig. 3-4. All measurements are expressed in inches.

Fig. 3-2. The 100th to 1/16th scale shown in the upper right-hand corner. Courtesy of Stanley Tools.

SIMILAR RIGHT TRIANGLES

In similar right triangles, the corresponding angles are equal and the corresponding sides are in proportion. This is the principle of similar right triangles. Shown in Fig. 3-5 are similar right triangles. Notice that the altitude and base have been doubled (increased in proportion) with a proportionate increase of the hypotenuse to 26.82; the angles remain the same. Conversely, if the measurements are reduced proportionately the sides will be reduced, but the angles remain the same. This is the underlying principle of laying out and cutting roof rafters using the framing square.

THE PYTHAGOREAN THEOREM

The Pythagorean theorem states that the hypotenuse squared is

Fig. 3-3. The location and description of the six tables used in roof framing. Courtesy of Stanley Tools.

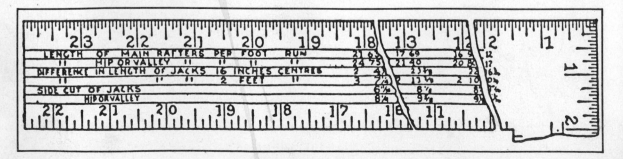

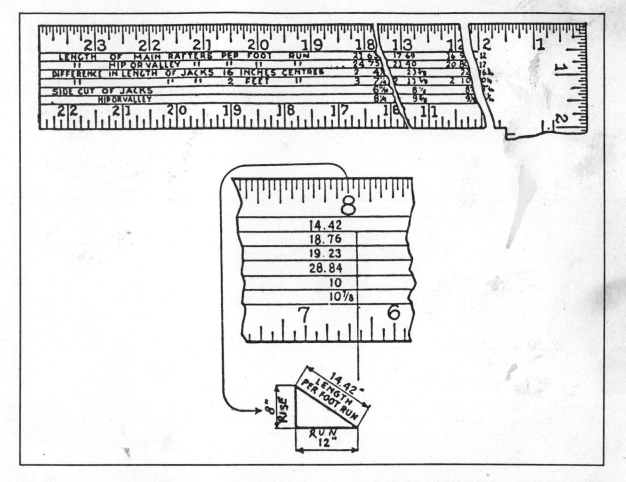

LENGTH OF MAIN RAFTERS PER FOOT RUN				21 63	17 69	16 97	12
" HIP OR VALLEY " " "				23 75	21 40	20 81	17
DIFFERENCE IN LENGTH OF JACKS 16 INCHES CENTRES				3 4⅓	2 3½	2⅝	6⅛
" " " 2 FEET "				3 7⅝	2 11½	2 10	9½
SIDE CUT OF JACKS				6⅝	8½	8½	
HIP OR VALLEY				8¼	9¼	9½	

| 14.42 |
| 18.76 |
| 19.23 |
| 28.84 |
| 10 |
| 10⅛ |

Fig. 3-4. The per-foot run of a rafter with an 8-inch unit of rise found on the first line of the framing tables.

equal to the sum of the sides squared. This theorem was developed, proven, and used by the Greeks thousands of years ago. It is one of the properties of the right triangle, and the most important mathematical principle in roof framing. See Fig. 3-6.

The Pythagorean theorem written as a formula appears as follows: The side C squared equals the sum of the sides A squared B squared, $C^2 = A^2 + B^2$.

In Fig. 3-6, observe that when side C is squared (multiplied by itself) the product is 25 square feet, which is the sum of the side A squared (16 square feet) plus the sum of the side B squared (9 square feet).

To get the final answer, the square root of 25 must be found, which is 5 (5 × 5 = 25 or 5 squared). In actual use, the formula is first transposed before any numbers are substituted, and from here on when the Pythagorean theorem is used the formula will be written as follows, $C = \sqrt{A^2 + B^2}$ expressed as side c equals the square root of the sum of the side a^2 and b^2.

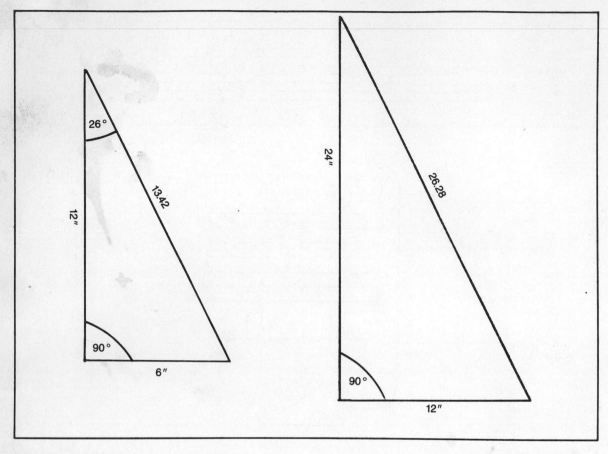

Example of the Pythagorean theorem:

$$C = \sqrt{3^2 + 4^2} = \sqrt{9 + 16} = \sqrt{25} = 5 \text{ feet}$$

The Pythagorean theorem is the mathematical process used to compute those figures shown on the framing table, and can be used to calculate all rafter lengths. The various methods of calculating rafter lengths are discussed in Chapter 6.

In computing rafter problems, always use the same units of measurements—either all feet or all inches. I also suggest that you use arithmetic to make your computations and then check the math with a calculator. Doing this will increase your proficiency in computing square root, and not make you a slave of the calculator.

LAYING OUT ANGLES WITH THE SQUARE

In carpentry, often you will be called on to lay out angles other than the right (90-degree) angle. The square in combination with the proper figures can simplify this job. To lay out a 45-degree angle with the square (Fig. 3-7):

Fig. 3-5. Comparison of similar right triangles. Courtesy of Stanley Tools.

Fig. 3-6. Right: The Pythagorean theorem shown graphically. Courtesy of Stanley Tools.

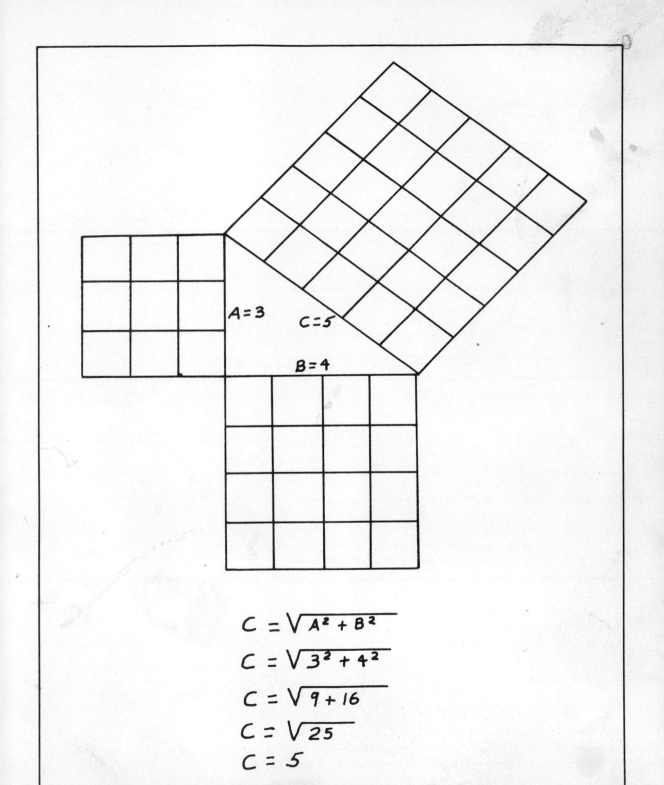

$$C = \sqrt{A^2 + B^2}$$
$$C = \sqrt{3^2 + 4^2}$$
$$C = \sqrt{9 + 16}$$
$$C = \sqrt{25}$$
$$C = 5$$

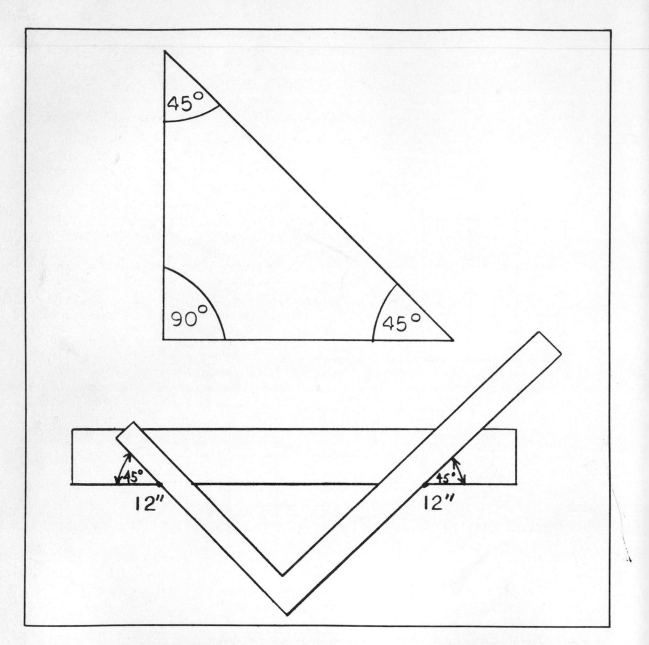

a. Lay the square on the edge of a straight board.

b. Choose a number on the edge of the square (any number will do). In this example, 12 will be the number.

c. Adjust the square so that 12 on the tongue and 12 on the body are exactly on the edge of the board. An angle of 45 degrees is formed with the two legs of the square as the sides and the edge of the board as the hypotenuse. Marking either leg will define a 45-degree angle in relation to the edge of the board.

Fig. 3-7. Using the square to lay out a 45-degree angle. Courtesy of Stanley Tools.

d. Important points to remember are that both numbers used must be on the same edge of the square (either inside or outside edges); they cannot be interchanged. When placing the square on the board, both legs must be placed on the same edge.

LAYING OUT ANGLES OTHER THAN 45 DEGREES

To lay out an angle of 30 degrees (Fig. 3-8):

a. In this example use 12 and 6 15/16.

b. Place 12 on the tongue and 6 15/16 on the blade. Mark the cut on the tongue. This will give an angle of 30 degrees.

c. Note that the complement of 30 degrees is marked on the blade of the square, giving an angle of 60 degrees. It is of equal importance to note that these angles were obtained using a horizontal plane as the reference line.

d. Refer to Fig. 3D-9 (page 177) to help clarify the difference between using a vertical line and a horizontal line as the reference line. Using a vertical reference line, the 30-degree angle is obtained by placing 12 inches on the tongue in a horizontal position and plac-

Fig. 3-8. Using the square to lay out a 30-degree angle. Courtesy of Stanley Tools.

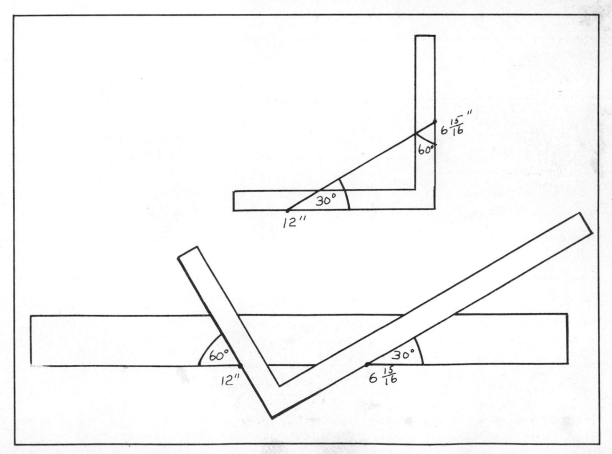

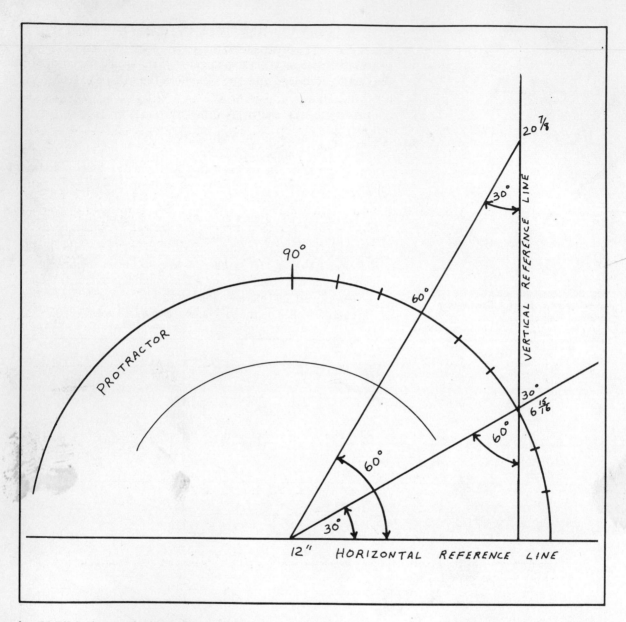

Fig. 3-9. Laying out the same angles
relative to a horizontal and vertical
base line.

ing 20 7/8 inches on the blade in a vertical position. Using a horizontal reference line, the 30-degree angle is obtained by placing 12 inches on the tongue and 6 15/16 on the blade. One angle is read using the blade and the hypotenuse, and the other is obtained by using the tongue and the hypotenuse.

It is important to note on what part of the square to mark the desired angle. If it is marked on the wrong leg of the square, the complement of the desired angle will be obtained.

THE 12TH SCALE

On the back of the square located on the outside edges of the tongue and blade are two scales 16 inches long on the tongue and 24 inches long on the blade, divided into increments of 1 inch, which in turn are divided into twelve parts or (12ths). With the square graduated in this form, 1 inch can be used to represent 1 foot and each 1/12 representing 1 inch. The scale can be reduced further where 1/2 inch or 1/4 inch equals a foot. Figure 3-10 shows the square placed on the edge of a board with 12 inches on each leg.

Mark the edges with a sharp pencil or steel scriber, and then measure the distance across the points (the hypotenuse of the right triangle) to get a reading of 17 inches. If the Pythagorean theorem is applied, the resulting answer would be 16.97 rounded off to 17 inches, proving that with care an accurate measurement can be produced with the scaling method.

Applying the same principle, using measurements of 5 and 15 (in units of feet), the measured result using the 1/12 scale is 15 feet 9 3/4 inches. See Fig. 3-11. Once again it can be checked by applying the Pythagorean theorem. The practical use of this principle is shown later in the book.

Fig. 3-10. Using the 12th scale on the outside edges of the back of the square to obtain the scale (bridge) length of a 45-degree angle. Courtesy of Stanley Tools.

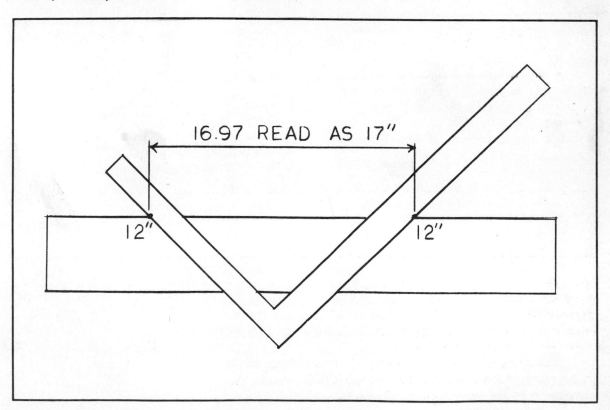

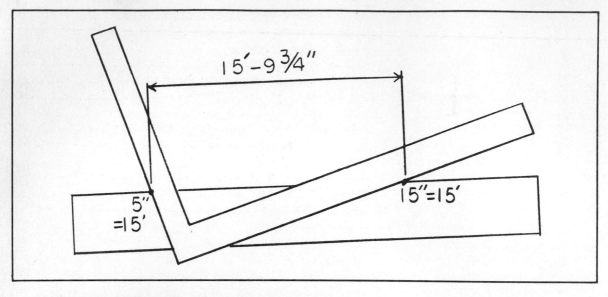

Fig. 3-11. Using the 12th scale on the outside edges of the back of the square to obtain the scale (bridge) length of a rafter. Note that the scale used is 1 inch equals 1 foot. Courtesy of Stanley Tools.

PRACTICE PROBLEMS AND PROGRESS CHECK

1. Using the 100th scale, draw a line 5.65 inches long. What is the measurement in inches and fractions of an inch? _____

2. What is the per-foot run for a rafter with a cut of 6 inches? _____

3. In a triangle with an altitude of 6 inches and a base of 12 inches, what is the length of the hypotenuse? _____

4. In a triangle with an altitude of 10 inches and a base of 12 inches, what is the length of the hypotenuse? _____

5. Using your framing square, lay out an angle of 36 degrees. What is the length of the hypotenuse? _____ What is the complement of the measured angle? _____

6. Using the 1/12 scale on the square, determine the length of hypotenuse of a triangle with an altitude of 10 inches and a base 24 inches. _____

7. Name the three parts of the steel framing square. a. _____, b. _____, and c. _____

8. Where are the roof framing tables located on the framing square? On the _____ of the _____

9. The figures on which line is used to calculate the length of a common rafter? _____

10. What are the figures in question 3 called? _____ _____ _____ of the _____ _____

11. Name the sides of the unit roof-framing triangle that are equivalent to the sides of a right triangle.

    a. Altitude = _____ _____

b. Base = _____ _____ .

c. Hypotenuse = _____ _____
_____

12. State the Pythagorean theorem _____

13. Using the framing square relative to the base line, what figures would be used to lay out a 30-degree angle? _____ and _____

14. Using the framing square relative to a vertical line, what figures would be used to lay out an angle of 30 degrees? _____

15. Describe how to lay out a 45-degree angle with the framing square. Do not forget to list the two most important points that have to be observed in performing this operation.

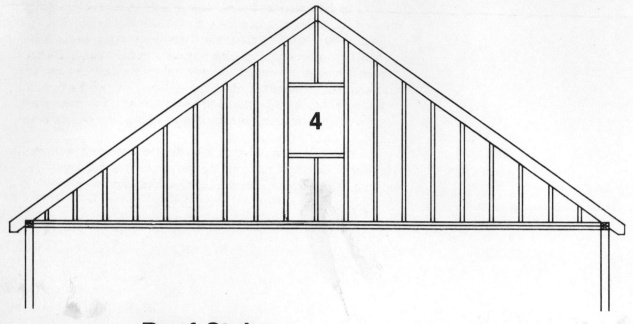

Roof Styles

Objective. Upon completion of this chapter, you should be able to do the following:

☐ Identify the basic roof types.
☐ Name all members that form the roof structure.
☐ Be able to identify the location of various roof members and understand their relationship to each other.
☐ Have the ability to visualize the basic roof styles from drawings.

PURPOSE AND FUNCTION OF A ROOF

The primary functions of a roof are to protect the house from weather and to play a key role in the overall design and construction of the structure. A roof must meet all building codes and be constructed to the highest standards of craftsmanship. The roof is supported by the weight-bearing exterior walls and, when necessary, interior walls that will also bear weight. With good planning and proper design, the roof contributes to the style and aesthetic features of the house.

ROOF TYPES

The Flat Roof. The simplest and most basic type of roof is the flat roof (Fig. 4-1). The primary member of this roof is called a rafter. It extends from the outside walls in a horizontal direction, usually spanning the narrowest direction of the house. It can be given a slight pitch to aid in the drainage of water, but it is still considered a flat roof. It is most popular in the so-called row house associated with urban housing, even though it can be found in new construction where a low silhouette is desired. With a flat roof, not only does the rafter function as a roof member but in many cases it also acts as a ceiling joist to provide a nailing surface for the finished ceiling below.

The Shed Roof. A variation of the flat roof is called the shed roof (Fig. 4-2). It is similar to the flat roof but it has a very pronounced slope in one direction. It usually is used in the construc-

Fig. 4-1. Section and isometric view of a flat roof.

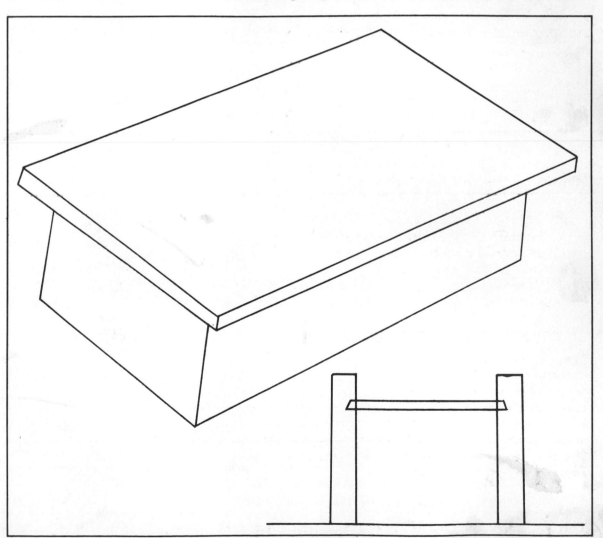

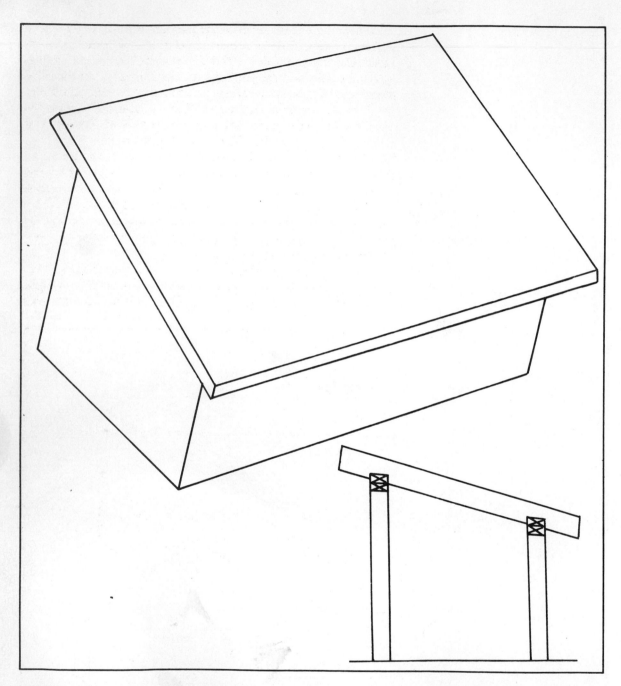

Fig. 4-2. Section and isometric view of a shed roof.

tion industry for sheds and temporary buildings. Another common use is for auxiliary buildings on farms. It is also very popular in contemporary construction where the use of clerestory windows is desired.

The Lean-to Roof. The lean-to-roof (Fig. 4-3) is the same basic structure as the shed roof, but the lean-to is usually associated with a structure that abuts another building. The lean-to roof can be attached to the building either at the wall or, if the building has a double-pitch roof, the shed can join it somewhere on the slope of the roof. This is the most popular style roof for additions to existing structures. It is also used in dormer construction either as a full dormer across the whole house or on a partial dormer. It is an extremely practical and easy roof to construct.

The Gable Roof. The gable (Fig. 4-4) or double-pitch roof is the basic roof from which the hip roof and intersecting hip and valley roofs are derived and combined to form most any other style roof. The gable is simple to calculate and easy to erect. It is relatively inexpensive to cover, easy to maintain, and practically trouble free for long periods of time. For these reasons it is one of the most popular roof styles.

The gable roof is composed of common rafters rising upward at an angle from the top of the outer walls to a point at the centerline of the house called the ridge. Extra strength is added to the roof

Fig. 4-3. Shed roof (lean-to) joining a building at a vertical wall.

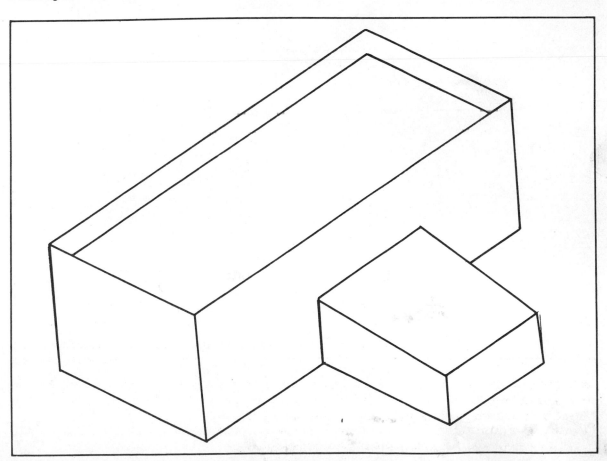

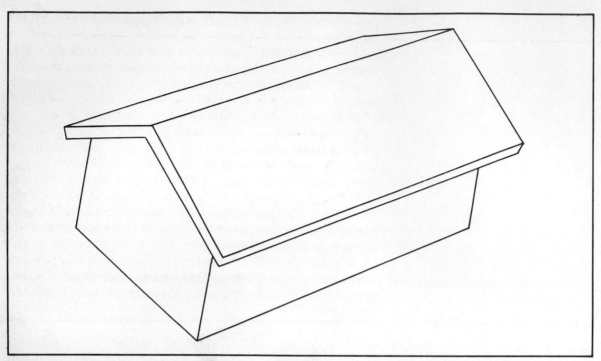

Fig. 4-4. The gable roof is a very popular and most common style. It is also called a double-pitched roof.

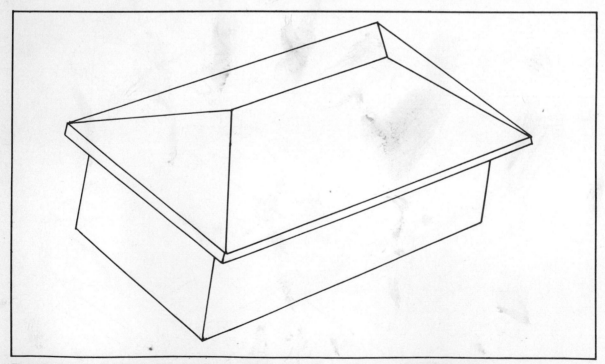

Fig. 4-5. A hip roof slopes in four directions, creating a protective overhang on all exterior walls.

with the addition of collar ties used in conjunction with the ceiling rafters.

The Hip Roof. The difference between a gable roof and hip roof (Fig. 4-5) is that the ends of a hip roof slope back toward the center of the house at an angle. These sloping lines created by the hip rafter tend to add a pleasing effect to the rather plain gable roof. It is a little more costly and more difficult to construct than the gable roof, but the aesthetic value the hip roof produces tends to increase its popularity.

Intersecting Valley Roof. When an addition or wing is added to a building, the intersection of these roofs creates a valley. From this configuration, the term valley roof (Fig. 4-6) is derived. If the ends are finished as a gable roof, it is called a gable valley roof. If the ends are constructed with hip rafters, they are called hip and valley intersecting roofs. You will find that the hip style ends are usually chosen when the builder wants to protect the building from the weather and sun by the creation of a large overhang. The overhang is made possible by the use of hip rafters.

Fig. 4-6. A gable and valley roof is created when two gable roofs intersect.

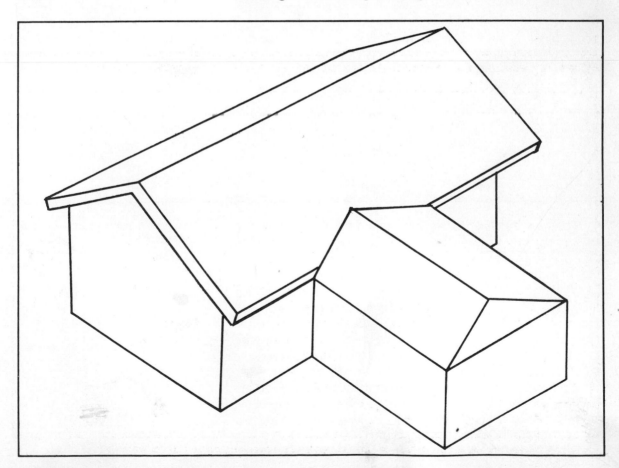

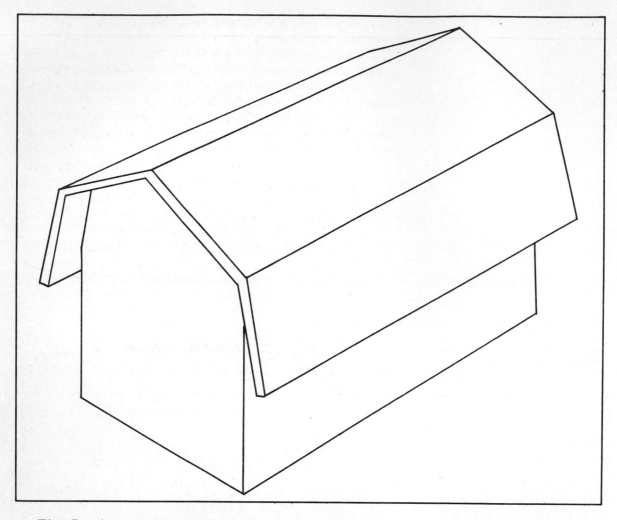

The Gambrel Roof. Another type roof derived from the gable roof is called the gambrel roof (Fig. 4-7). It has two pitches (instead of one as in the gable roof). The lower rafters have a greater pitch than the upper rafters. The gambrel roof is most popular when used in barnlike buildings. For this reason, it is sometimes called the "barn roof." In this type of construction, the roof can be well braced without obstructing the area in the center part of the building. This is a great asset in residential construction because the upper floor is expanded. The addition of windows increases the living space.

If the gambrel style has a drawback it is that some designers feel that it is awkward and does not enhance the appearance of the house. It has been shown that, with proper design, the gambrel roof can be made to look most attractive.

The Clerestory Roof. Clerestory often describes the type

Fig. 4-7. A gambrel roof is a a variation of the gable roof.

of windows that are being used more than the style of roof. The clerestory roof consists of two shed roofs. The roofs can have the same pitch or each can have a different pitch; the designer has the choice. This style is used to allow light and ventilation into a room when a cathedral ceiling is constructed. As shown in Fig. 4-8, one wall is higher than the wall of the roof that abuts it, allowing the installation of windows. These windows can be of fixed (with no means of opening or closing) or they can be hand operated or power operated, allowing them to open or close.

Gable Roof with a Dormer. Figure 4-9 shows a basic gable roof with a dormer built into it. The dormer has a shed roof. The dormer roof can be any style (depending on the style of the house and what effect the designer is trying to achieve). A small dormer of this type is usually used for light or ventilation.

An Intersecting Shed Roof. The type of roof shown in Fig. 4-10 is used when a wing abuts an existing structure. It intersects on one side of the double-pitched roof somewhere on the slope. It is often used when a porch, a portico, or a patio is added to an existing structure.

BASIC ROOF-FRAMING MEMBERS

Figure 4-11 shows the location of all the roof framing members on

Fig. 4-8. The clerestory roof consists of two separate shed roofs, making it possible to install windows in one wall.

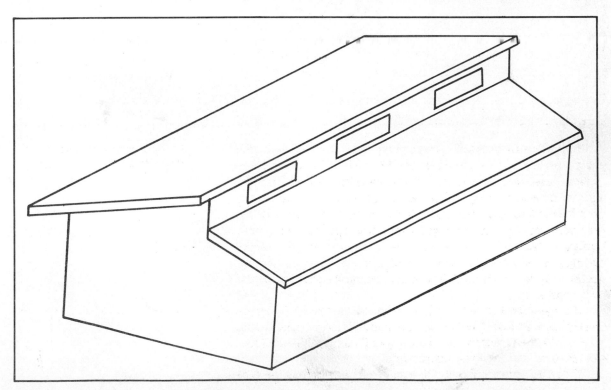

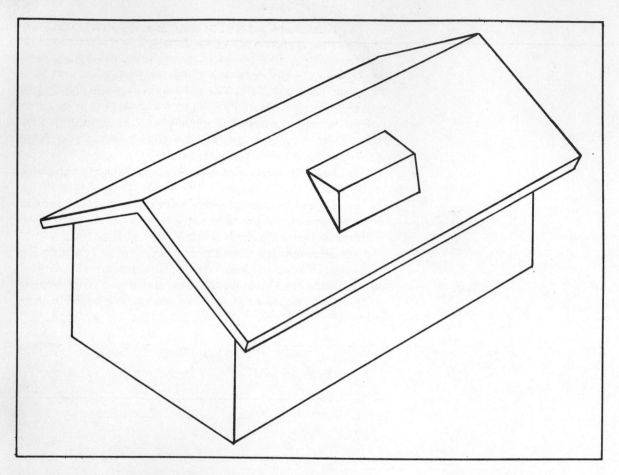

an equal pitch, unequal span, intersecting hip and valley roof. Descriptions of all the roof framing members and how to recognize them follow.

Fig. 4-9. A gable roof with a small shed dormer.

Top Wall Plates. The uppermost horizontal members of the outside walls consist of two units, called bottom and top plates, fastened together and so designed to bear the load placed upon them by the roof structure.

Ridge Board. The ridge board is a horizontal member that runs the full length of the roof. Forming the uppermost portion of the roof, the ridge board is where the two opposite rafters meet at the peak. Structurally it does not give any significant strength to the roof. Its prime function is to make alignment and placement of the rafters easier. There are many roofs in existence that were built and are being built without ridgeboards.

Major Ridge. The major ridge is associated with the main section of the structure.

Minor Ridge. The minor ridge is associated with wing additions to the structure.

The Common Rafter. The common rafter runs from the wall plates to the ridge at right angles to both.

Hip Rafter. The hip rafter runs from the outside corner of the top plates to the ridge at a 45-degree angle.

Long Valley Rafter. The valley rafter runs from the inside corner of the top plates to the ridge at a 45-degree angle.

Short Valley Rafter. The short valley rafter runs from an inside corner of the top plates and intersects the long valley rafter at a point on the rafter at 45 degrees.

Hip Jack Rafters. Hip jack rafters run from the wall plates to a hip rafter at right angles to the plates.

Valley Jack Rafters. Valley jack rafters run from the ridge to a valley rafter at right angles to the ridge.

Hip-to-Valley Cripple Jack Rafters. Hip-to-valley cripple jack rafters run from a hip rafter to a valley rafter at right angles to the ridge and valley, but do not touch them.

Valley-to-Valley Cripple Jack Rafters. Valley-to-valley cripple jack rafters run from a valley rafter to a valley rafter at right angles to the ridge and wall plates, but do not touch them.

Fig. 4-10. A gable roof with an intersecting shed roof. The point of intersection can be anywhere on the slope.

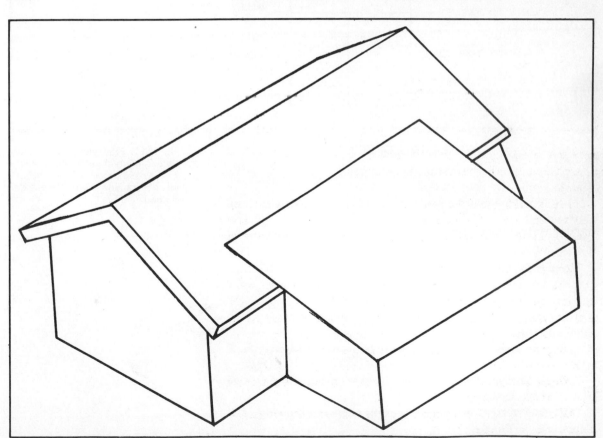

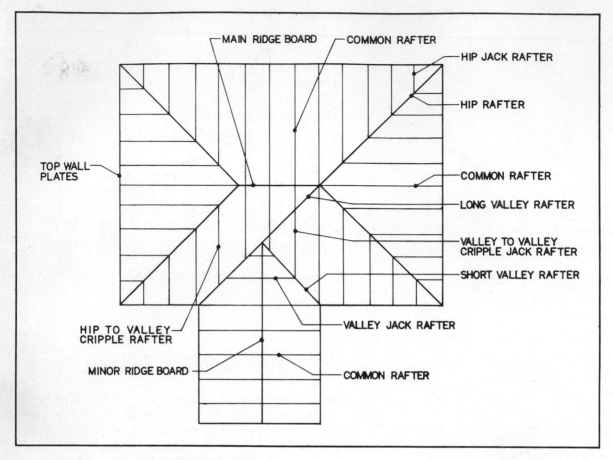

MAIN RIDGE BOARD — COMMON RAFTER

HIP JACK RAFTER

HIP RAFTER

TOP WALL PLATES

COMMON RAFTER

LONG VALLEY RAFTER

VALLEY TO VALLEY CRIPPLE JACK RAFTER

SHORT VALLEY RAFTER

VALLEY JACK RAFTER

HIP TO VALLEY CRIPPLE RAFTER

MINOR RIDGE BOARD

COMMON RAFTER

PROGRESS CHECK

1. What is the basic function of a roof?

2. Name five roof styles.

3. Is a ridge board absolutely necessary in the construction of a roof?

4. What is the material called that covers the roof rafters?

5. Wood boards are the most popular material in use today to cover roof rafters. T — F.

6. The common rafter touches the plates of the outer walls and the ridge. T — F.

7. The common rafter is placed at a 45-degree angle to the ridge and plates. T — F.

8. The lean-to and shed roofs are basically the same type. T — F.

9. Are hip and valley rafters placed at an angle to the ridge board? Yes — No.

10. Are jack rafters placed at right angles to the ridge and plates? Yes — No.

Fig. 4-11. Plan view of an intersecting hip and gable roof forming a valley at the point of intersection. All roof framing members are described.

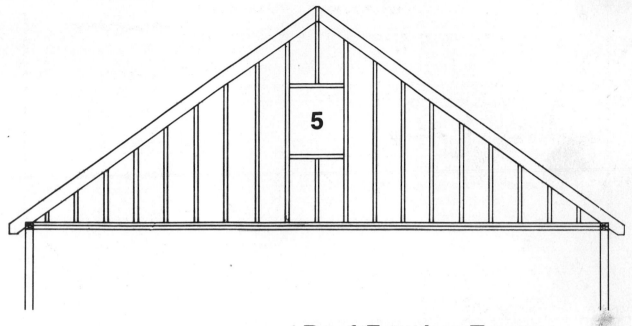

Roof Framing Terms

Objective. Upon completion of this chapter, you should fully understand the following terms, their relationships and how the terms are used in calculations for and construction of a roof.

- ☐ Total span.
- ☐ Total run.
- ☐ Total rise.
- ☐ Unit span.
- ☐ Unit run.
- ☐ Unit rise.
- ☐ Angle, slope, cut, and pitch.
- ☐ Line length.
- ☐ Plumb line and level line.
- ☐ Inclination.

Prior to starting this unit, it is suggested that you review, understand, and can perform the following mathematical operations.

- ☐ Addition of decimals.
- ☐ Subtraction of decimals.
- ☐ Multiplication of decimals.
- ☐ Division of decimals.
- ☐ Conversion of decimals to fractions of a foot or fractions of an inch.
- ☐ Conversion of fractions to decimals of a foot or decimals of an inch.

☐ Square root calculations.
☐ Pythagorean theory.
☐ Similar right triangles.

In the study of roof framing, you must become familiar with a number of roof framing terms. It is of the utmost importance that you study the following terms and that you understand them.

Span. Span is the distance between the outside walls of a structure from the outside edge of one plate to the outside edge of the opposite plate in a horizontal plane. The span distance is shown in the plans of the house. This measurement is usually shown on the floor plan or the roof framing plan if it is included with the complete house plans. The span can be stated as twice the run.

Run. One-half the span is called the run (also referred to as "total run"). It is the measured distance from the center of the span to the exterior walls of the structure. See Fig. 5-1.

Rise. The vertical distance the ridge rises above the plates at the centerline of the span is called the rise (Fig. 5-1). It is also referred to as "total rise." The above terms are used when refer-

Fig. 5-1. The three basic terms used in roof framing calculation.

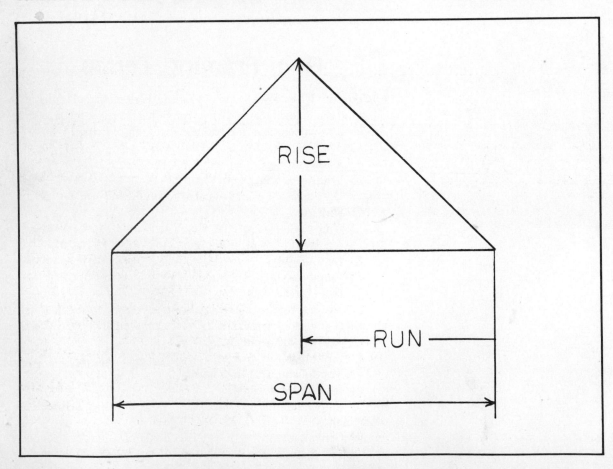

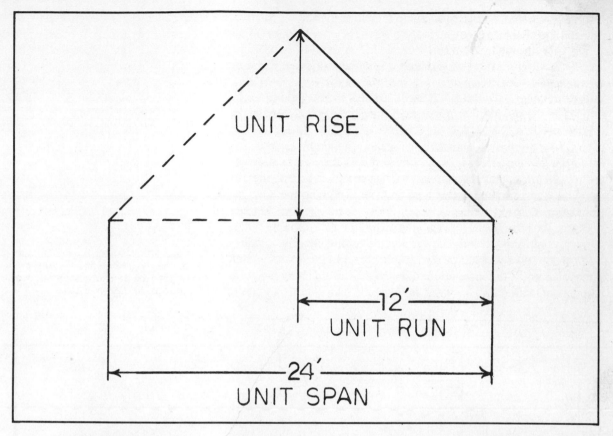

UNIT RISE

12′
UNIT RUN

24′
UNIT SPAN

Fig. 5-2. The three basic terms expressed in "units." Note that the unit run and unit span are fixed measurements. Unit rise (x) can and does vary.

ring to the complete roof or the roof in full size. To develop the full-size roof, use a similar triangle that has been reduced in size and expressed as "units." With the use of these units (given in inches) and the steel framing square, any style and size of roof can be developed.

Unit of Run. The unit or run (Fig. 5-2) is a fixed measurement; it is always 12 inches or 1 foot. This measurement was chosen because it is a standard unit of measurement in the construction industry. Any measurement in a horizontal direction is expressed as run, and is always taken in a level plane.

Unit of Rise. The number of inches a roof rises for every unit of run (foot) is called the unit of rise (Fig. 5-2). This value is always given in inches. It can and does vary with each roof. Unit of Rise is usually given in the elevation drawings of the plans. It is shown by a drawing of a small right triangle.

Unit of Span. If the total run is one-half the span, it can also be stated that span is twice the run. We can then assume that the unit of span is always 24 inches or twice the unit of run of 12 inches. See Fig. 5-2.

Plumb and Level. A plumb line is any measurement or cut

that is made in a vertical direction. This line and cut is made on the tongue of the square and is the rise of the rafter. A level line is any measurement or cut that is made in a horizontal plane. This line and cut is marked on the body of the square and is the run of the rafter. See Fig. 5-3.

Remember, all horizontal measurements are made at right angles to the plumb line. All vertical measurements are made on the plumb line.

To successfully construct a roof as envisioned by architect or designer, the carpenter must be supplied with the proper information. This information is conveyed to the carpenter through the use of plans that contain drawings, dimensions, and other detailed information. If the structure is to be a flat roof, a carpenter only has to know the span in order to find the rafter length and cut it. When a more complicated roof—such as a gable roof, a hip roof, a valley roof or any combination of these styles are to built—a carpenter must also know the inclination of the roof. This inclination can be shown and stated in four ways: as the angle, slope, cut, and pitch.

Fig. 5-3. A framing square is used to lay out a rafter and the relationship of the parts of a square to the lines being drawn.

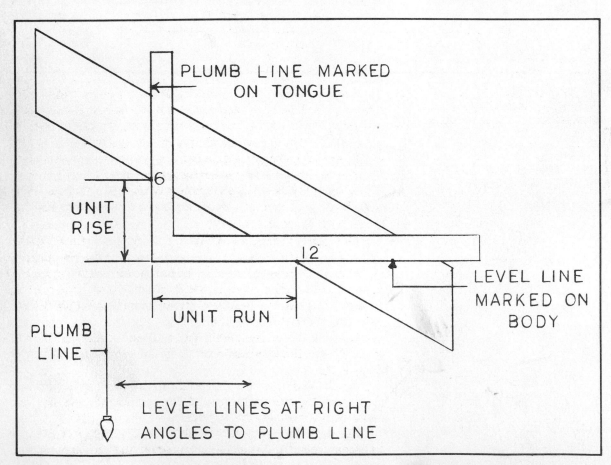

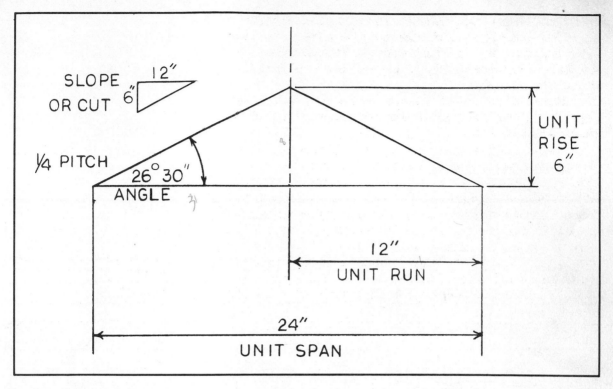

Fig. 5-4. Comparison of roof framing terms expressed in units shown graphically.

Angle. Angle (Fig. 5-4) is the inclination a rafter makes with a horizontal plane at the top plates of a building. It is expressed in degrees and minutes. In the United States, the use of angle is not a common means of expressing the inclination.

Slope. Slope (Fig. 5-4) is expressed in inches of rise for every foot run. It can be expressed as inches and foot run written as (6 and 12) or graphically shown by a small triangle called the slope triangle. You must bear in mind that the unit of rise will vary but the unit of run does not change.

Cut. Cut (Fig. 5-4) is another term to express slope. It has the same meaning and is also shown in the same fashion as slope.

Pitch. Pitch (Fig. 5-4) is another term used by the craftsman to express the inclination of a roof. It is the ratio of the rise to the span expressed as fraction. Some of the more popular pitches are 1/4, 1/2, 1/3, and 1/6.

Compare the terms shown in Fig. 5-4, and become familiar with them. Make sure you understand how the terms are used interchangeably.

Figure 5-5 shows how the rise is proportional to the span. It's the ratio of the rise to the span. When calculated in full size, the measurements are always in feet.

Figure 5-6 shows a comparison between unit measurements and full measurements. In unit measurements, the run and span are

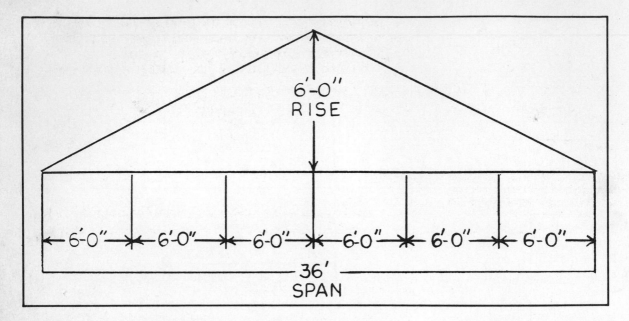

Fig. 5-5. The rise is proportional to the span. In this example it is a 1/4 pitch.

always 12 inches and 24 inches. The rise can and does vary. When full dimensions are given, all three measurements do vary (depending on the dimensions and design of the roof).

With the use of the framing square as shown in Fig. 5-7, the relationship between slope and pitch is demonstrated. Notice that with a rise of 18 inches and a span of 24 inches, a 3/4 pitch is obtained.

$$\frac{18}{24} = \frac{3}{4} \text{ when reduced}$$

For practice, work out the other pitches. Keep in mind that all the measurements are in inches (units).

A summary of all the roof framing terms and their relationship is shown in Fig. 5-8.

FORMULAS AND PRACTICE PROBLEMS FOR ROOF FRAMING

A. To find total rise when unit rise and total run are given, multiply unit rise by total run.

Unit Rise × Total Run = Total Rise

Where:
Unit of rise is given in inches.
Total run is in feet.
Answer: given in inches.

Example. Given a unit of rise of 6 inches and a total run of 8 feet, what is the total rise? Applying the above formula, multiply 6 inches by 8 feet for the unit rise: $6 \times 8 = 48''$.

Given the following information, calculate the total rise.

UNIT RISE	TOTAL RUN	ANSWER
3″	15′ 0″	
8″	12′ 0″	
6 1/2″	14′ 3″	
8 1/4″	15′ 8″	

Fig. 5-6. The proportionate relationship between the pitch found using units of measurements as opposed to measurements in feet.

Note: It is suggested that 6 1/2″ and 6 1/4″ be changed to inches

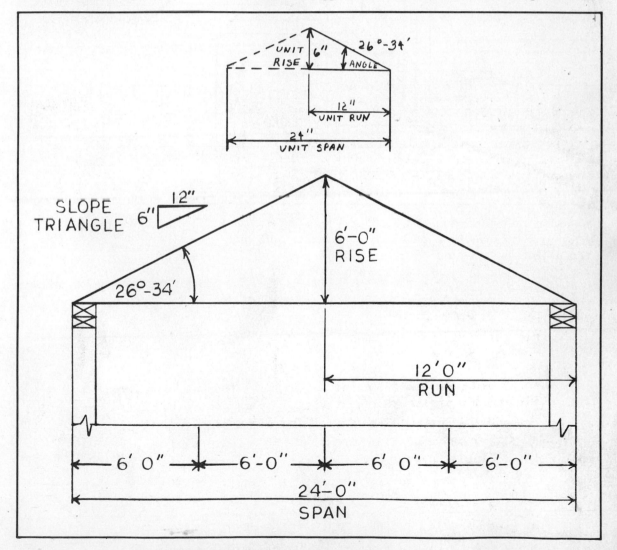

87

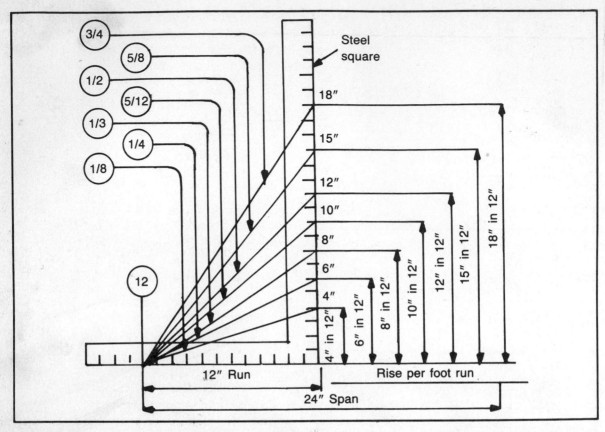

and decimals of an inch and, that 14′ 3″ and 15′ 9″ be changed to feet and decimals of a foot before proceeding with any calculations.

Fig. 5-7. The relationship between slope and pitch with the use of the framing square.

B. To find unit rise when total rise and total run are given, divide total rise by total run.

Total Rise ÷ Total Run = Unit Rise

Where:
Total rise is in inches.
Total run in feet.
Answer: unit rise in inches.

Example. Given a total rise of 72 inches and a run of 12 feet, find the unit rise in inches: 72″ divided by 12′ equals 6″

Given the following information, calculate the unit rise.

RISE	RUN	ANSWER
96″	18′ 0″	
3′ 6″	14′ 6″	
48″	12′ 0″	
5′ 0″	15′ 0″	

Figures in the above problems must be changed to the proper units of measurement before proceeding with any calculations. Any decimals in the answer must be changed to a fraction to the nearest 1/16th of an inch.

C. To find pitch when total rise and total span are given, remember that pitch is the ratio of rise to span, stated as a fraction of total rise over total span.

$$\frac{\text{Total Rise}}{\text{Total Span}} = \text{Pitch}$$

Where:
Total rise is in feet.
Total span is in feet.
Answer: pitch is stated as a fraction.

Given the following information, a total rise of 8 feet and total span of 32 feet, find the total pitch:

$$\frac{8'}{32'} = \frac{1}{4} \qquad \frac{2}{7}$$

Fig. 5-8. Summary and comparison of the roof framing terms. There is a direct relationship between unit measurements and total measurements.

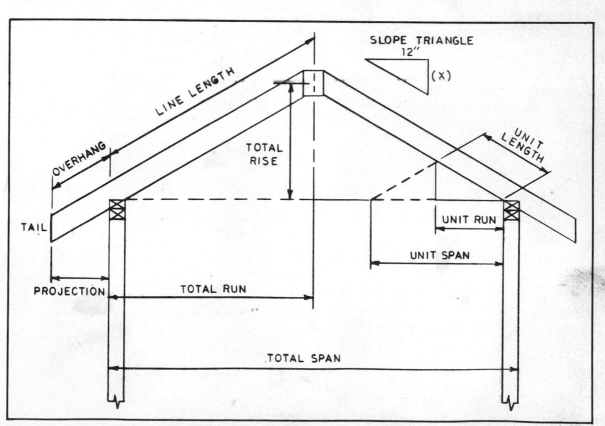

TOTAL RISE	TOTAL SPAN	ANSWER
5' 0"	20' 0"	
4' 0"	36' 0"	
84"	21' 0"	
18' 0"	24' 0"	

Note: Change all figures to the proper units of measurements before proceeding with any calculations.

D. Finding the unit rise or cut of a roof, when pitch and total span are given, is a two-step procedure.

1. Multiply the pitch by the span to find the rise which will be in feet.

2. Divide the total rise by the total run. This will give the unit rise in inches: (Pitch × Total Span) ÷ Total Run = Unit Rise.

Where:
Pitch is shown as a fraction.
Total span is in feet.
Total rise changed to inches.
Answer: unit rise will be in inches.

Given the following information, a 1/3 pitch and a total span of 24 feet, calculate the unit rise in inches.

1/3 × 24' = 8' is the total rise.

Multiply 8' by 12" to get a total rise of 96". Divide the total rise 96" by the run of 12' to get the unit rise of 8".

Note: To find the total run, take 1/2 the span of 24' to get a total run of 12'.

Practice Problems:

PITCH	SPAN	ANSWER
1/6	36' 0"	
1/8	32' 0"	
1/2	28' 0"	
1/3	27' 0"	

PROGRESS CHECK

1. The run is always twice the span? T — F

2. The unit of run is always _____ inches.

3. The vertical distance the ridge is above the plate is called the _____ .

4. The unit of span is always _____ inches or _____ the unit of run.

5. The unit or rise is always a fixed measurement. T — F

6. A line drawn in a horizontal direction is called a _____ line.

7. A line drawn in a vertical direction is called a _____ line.

8. The inclination a rafter makes with the top plates of a building is an _____ and expressed in degrees.

9. Pitch is always shown as a _____ .

10. Slope and cut are both shown and expressed in the same way. T — F.

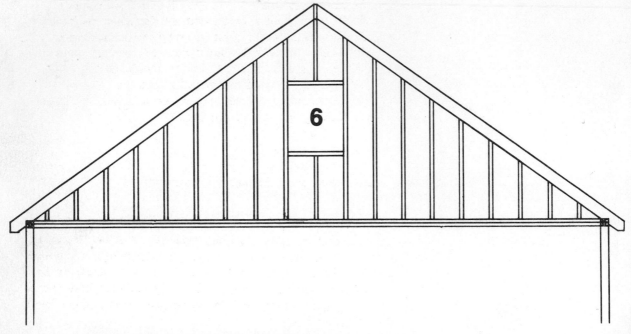

6

The Common Rafter

Objective. In this chapter you will learn how to calculate, lay out, and cut a common rafter. As the name implies, the common rafter is the basic rafter used in conventional roof framing. The figures used to calculate all other rafters are those of the common rafter. The building of a roof progresses from a simple operation for a gable or a shed roof, to a more complex roof such as a hip or a hip-and-valley intersecting roof. When building a gable roof, the experienced carpenter usually will not find it necessary to refer to a roof-framing plan. As the roof design becomes more complex, however, a roof-framing plan must be referred to.

To prepare you for the construction of more complex roofs and also to make the job of erecting a gable roof easier, this chapter also covers the fundamentals of drawing a roof framing plan. From this plan, you will be able to visualize the layout of the rafters, determine the number of rafters needed to build the roof, measure the run of the rafters, and determine the projection of the overhang and the length of the ridge board. In conjunction with a section view of the roof, the length of the rafter, the length of the overhang, and the angle the roof makes with the horizontal plane can all be obtained.

Upon completion of this chapter you should fully understand the following.

☐ How to draw a roof-framing plan.
☐ How to calculate the length of a common rafter.
☐ How to lay out a common rafter using the step-off method.
☐ How to lay out a common rafter by measurement.
☐ Steps in cutting the bird's-mouth.
☐ Marking and making the ridge plumb cut.
☐ Marking and making the tail plumb cut.
☐ Calculating and laying out the projection.
☐ Making and using a story pole.
☐ Erecting the roof.
☐ Collar beams.
☐ Gable studs.
☐ Framing an opening in a gable wall.

THE ROOF-FRAMING PLAN

The use of a roof-framing plan simplifies the erection of any roof. If you are using a set of professional drawings, it is not uncommon for the draftsman to include a roof-framing plan from which all measurements can be taken. If you do not have professional plans, it is best to make a drawing before you proceed with the actual work.

A roof-framing plan is usually drawn to scale. The scale used depends on the size of the roof to be built and size of paper used. The larger the scale (within reason) the more accurate the drawing will be. It is suggested that a scale of 1/4 inch or 1/2 inch to the foot be used. When an architect's scale is not available, the standard 1-foot rule can be used. If a 1/4 scale is chosen, there will be 4 feet to the inch. If a 1/2 scale is used there will be 2 feet to the inch. The comparison of line length between 1/4 and 1/2 scale is shown in Fig. 6-1.

Drawing A Gable Roof-Framing Plan

a. Draw an outline of the roof to the chosen scale. Use a dashed line to indicate the outside of the wall plates. See Fig. 6-2.

b. Draw a solid line paralleling the dashed lines (wall plate line). This outlines the projection of the tail piece.

c. Find the center of the span and draw the ridge. If there is no overhang (projection) at the ends of the house, the ridge is the same dimension as the length of the house.

d. Mark the position (layout) rafters of the 16 inches on center (o.c.), starting from either end of the building. I like to do any layout from left to right, looking at the building from the front. Make sure both front and back walls are laid out from the same end, and in the same direction.

e. A section view (Fig. 6-3) will help determine the length of

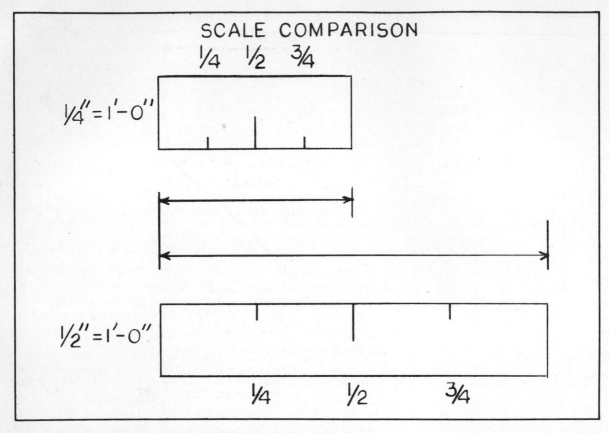

the rafter stock needed for the rafters. It will also help to determine the length of the overhang and the angle of the roof, if desired. With the aid of the plan and elevation views, the completed roof now can be visualized.

Fig. 6-1. The difference in length of one foot of measurement using two scales. It can also represent inches.

Basic Data Needed to Construct a Roof

To build a roof, a carpenter must know the span of the building. From this the run is determined and the angle of inclination is found, stated as the pitch. The cut of the roof is shown in the plans as a slope triangle (Fig. 6-3). Without the preceding information and a clear understanding as to their relationships, it will be most difficult to build a roof properly. The two phases in the actual construction of a roof are laying out and cutting of the rafters and the erection of the roof.

LAYING OUT THE COMMON RAFTER

The parts of a common rafter are shown in Fig. 6-4.

a. The theoretical length of a common rafter is calculated mathematically. It is also called the "line length" or "measured

Fig. 6-2. Right: The steps involved in drawing a basic roof-framing plan.

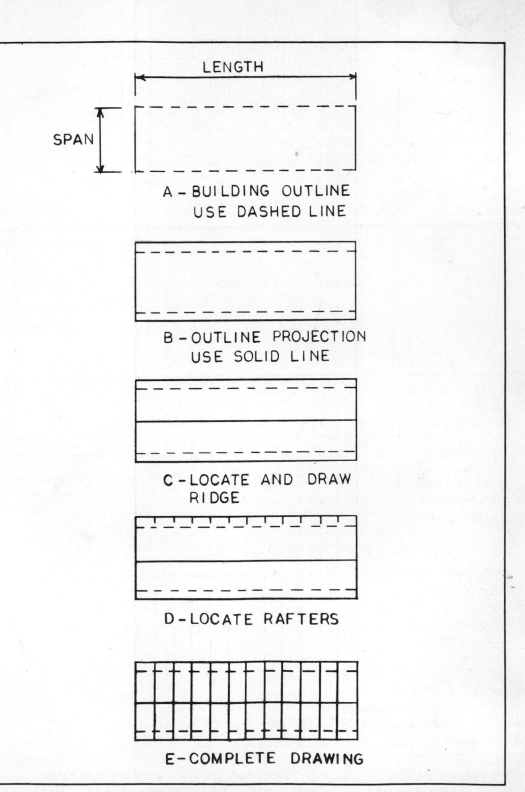

LENGTH

SPAN

A – BUILDING OUTLINE
USE DASHED LINE

B – OUTLINE PROJECTION
USE SOLID LINE

C – LOCATE AND DRAW
RIDGE

D – LOCATE RAFTERS

E – COMPLETE DRAWING

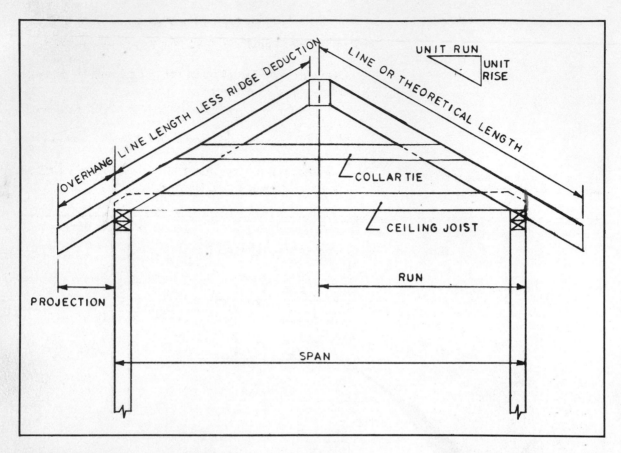

length" of the rafter. It is measured along either edge or on the rafter measuring line.

b. The overhang of the rafter is measured along the top edge of the rafter from heel line to tail cut.

c. The measuring line is an imaginary line extending from the intersection of the bird's-mouth heel cut and outside edge of the top wall plate parallel with the edges of the rafter to the ridge. It is marked on the rafter with the use of a chalk line.

d. The tail cut is the actual end cut of the rafter.

e. The tailpiece is another name for the overhang. This, in conjunction with the fascia trim and soffit, form the cornice of the roof.

f. Ridge plumb cut is the uppermost cut on the rafter. It is part of the rafter that fastens to the ridge board.

Bird's-mouth cut on the bottom edge of the rafter at the wall plates. The cut is made into the rafter to help spread the load of the roof on the rafter edge and help prevent the rafter stock from compressing, eventually causing a sag in the roof. Figure 6-5 shows the parts of a bird's-mouth.

Fig. 6-3. The information necessary to construct a roof can be determined from a section view such as this one shown.

Tail Cuts of a Rafter

Figure 6-6 shows some of the most popular tail cuts of a common rafter.

a. Overhang with the tail cut plumb. This results in the fascia installed in a vertical position.

b. The tail cut flush with no overhang results in the fascia being flush with the outside walls of the structure.

c. Overhang with the tail cut square results in the fascia trim sloping back toward the walls of the house.

d. Overhang with a plumb and level tail cut is always used when a boxed cornice is specified in the plans.

Fig. 6-4. Defines the names of the parts of a common rafter.

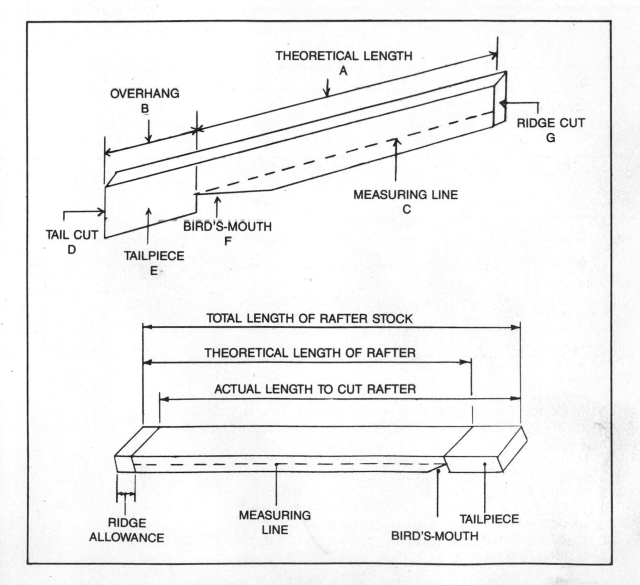

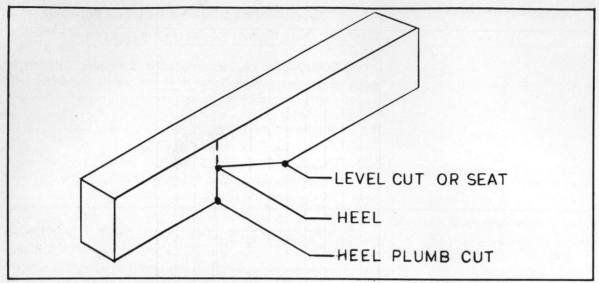

Fig. 6-5. Shows the three principle parts of a bird's-mouth.

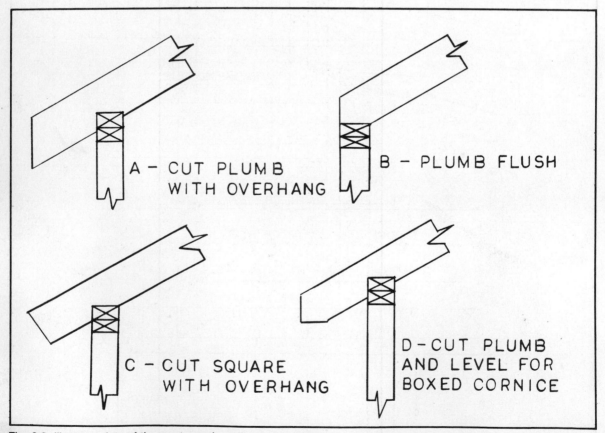

Fig. 6-6. Illustrates four of the most popular tail cuts on a rafter.

Steps In Laying Out the Common Rafter

Figure 6-7 shows the six steps involved in laying out the common rafter.

    a. Calculate the theoretical length of the rafter. This length is also called the measured length or line length of the rafter.

    b. Lay out the ridge cut.

    c. Lay out the bird's-mouth.

    d. Measure the overhang.

    e. Lay out the tail cut.

    f. Make the deduction for ridge shortening.

    These steps are described, in the text that follows, in the order that they are to be performed.

Calculating the Theoretical (line) Length of a Rafter

Scale Method. There are four ways to calculate the length of a rafter. Figure 6-8 demonstrates the use of the 1/12 scale to find the line length. This is called the bridge or scale method. In using this method, it is best to make all marks using a knife or steel scriber on a sharp edge of a board. You will find it is the least ac-

Fig. 6-7. The six steps in laying out a common rafter.

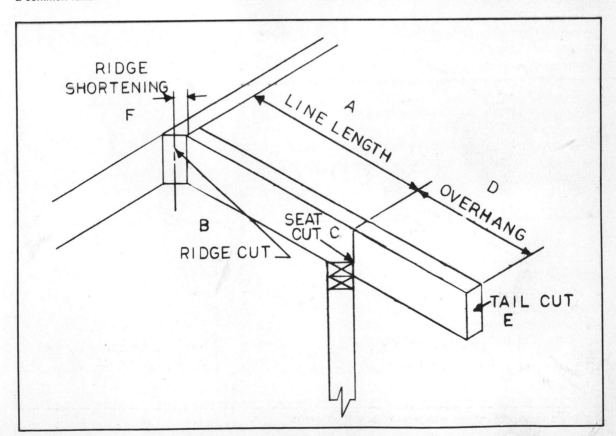

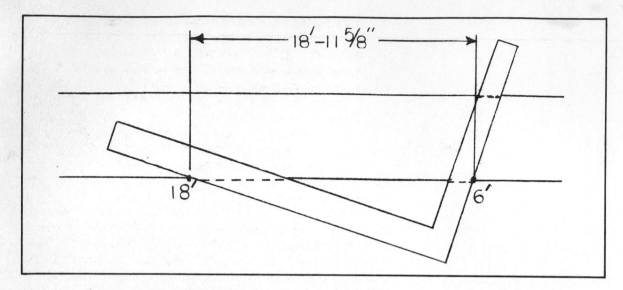

Fig. 6-8. Using the 1/12 scale on the square to determine the theoretical length of a common rafter.

curate method for finding the line length but it is a most practical means to estimate the length of stock needed for the rafter (remember to add the length of the overhang). In using this method notice that all dimensions are designated in feet. The total rise is marked on the tongue and the total run is marked on the body of the square. Using the body of the square, the bridge or scale distance is read on the outside edge of the square with the feet and inches read directly from the scale and the fraction of the inch approximated.

Step-Off Method. The step-off method has been used by carpenters for many years and still is popular. It is a method that requires no math to calculate the length and lay out of a rafter. Unless extreme care is used, errors can be introduced in the layout process. If a blunt pencil is used, the mark can be as much as 1/16 of an inch in width. Multiplying this 15 times for a run of 15 feet will introduce almost 15/16 of an inch (or almost one full inch) of error.

As an example in laying out a short rafter with a run of 4 feet and a unit rise of 8 inches, the following sequence of steps are performed. See Fig. 6-9.

1. Place the square on the edge of a board. Using the unit rise of 8 inches on the tongue and unit run of 12 inches on the body, draw a line on the tongue. This is the ridge plumb line. Mark the run with a small line or dot at the 12-inch graduation on the square. The measured length between the two points is 14 7/16 inches (14.42). This is the hypotenuse of the right triangle formed by the unit rise of 8 inches and the unit run of 12 inches, called the per foot run of the common rafter, and can be found on the first line of the framing table under 8 inches (14.42).

100

2. Move the tongue of the square to the 12-inch mark. Using the same cut of 8 and 12, make a new small mark at the 12 inch increment. Once again it measures 14 7/16 inches.

3. Repeat the same procedure used in steps 1 and 2.

4. Repeat the same procedure used in steps 1, 2, and 3.

5. Move the square to the last small mark and draw a line. This line is the heel cut.

Stepping off a rafter (Fig. 6-10) is accomplished by using the unit triangle with a rise of 8 inches and a run of 12 inches, moving the square four times, and drawing the heel plumb line on the fifth mark—creating four spaces between the ridge plumb line and the heel plumb line—and producing a right triangle A, B, C, of which the hypotenuse is the theoretical length of the rafter. Notice that the ridge and heel plumb lines are parallel. When the rafter is in place, these lines are in a vertical (plumb) position.

Line Length Determined Mathematically. Another method used to determine the length of a rafter is using the per foot run (stated in inches and found on the rafter table under the unit of rise figure for the particular roof). Multiply it by the total run stated in feet. This will give the line length in inches (which is changed to feet, inches, and fractions of an inch). An example,

Fig. 6-9. The use of unit measurements and the square to lay out a full-size rafter.

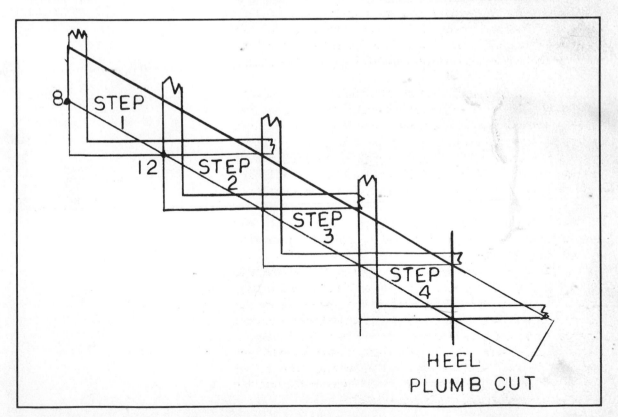

8

STEP 1

12   STEP 2

STEP 3

STEP 4

HEEL
PLUMB CUT

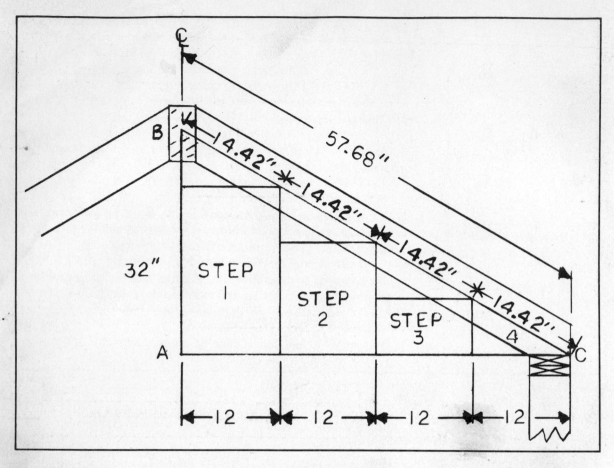

Fig. 6-10. Shows the development of the theoretical framing triangle of a full-size rafter.

using the measurements of 8 and 12 with a run of 4 feet, is worked out in Fig. 6-11.

The Pythagorean Theorem Used to Calculate Line Length. Rafter length can also be calculated using the Pythagorean theorem. It is important that all dimensions used be in the same units (either all feet or all inches). The results are converted to feet, inches, and fractions of an inch. A practical problem, using 8 and 12 as the slope with a run of 4 feet, is shown in Fig. 6-12. I suggest you use a calculator to compute the rafter length.

1. Obtain the rise using the formula: Rise = Unit Rise in Inches Multiplied by Total Run in Feet.

2. Change the answer in inches to feet.

3. Arrange the values for the Pythagorean theorem.

4. Using the calculator, perform the required functions.

5. Change the answer found in feet to feet and inches.

In comparing the mathematical answers, it is found that there is a slight difference between the Pythagorean theorem and the framing square table method of calculating the rafter length. In

102

STEP 1 - Find the line length

14.42 Per Ft Run
×   4 Run Of Rafter
————
57.68″ Line Length of Rafter

STEP 2 - Change inches to feet

$$\begin{array}{r} 4 \\ 12\overline{\smash{)}57.68} \\ \underline{48} \\ 9 \text{ remainder} \end{array}$$

STEP 3 - Convert decimal of an inch to a fraction

$$\begin{array}{r} .68 \\ \times\ 16 \\ \hline 408 \\ 108 \end{array}$$

$$14.88 = \frac{14.88}{16} = \frac{15}{16}$$

ANSWER

$$4' - 9\frac{15''}{16}$$

Fig. 6-11. An example in computing the line or theoretical length of a rafter using the table on line one of the square.

1. Find total rise: 4′ × 8″ = 32″.
2. CONVERT inches to feet: 32″ ÷ 12″ = 2.66′.
3. Arrange Pythagorean theorem:

$$\sqrt{2.66^2 + 4^2}$$

4. Find square root using calculator:

$$\sqrt{7.08 + 16} = \sqrt{23}$$

$$\sqrt{23} = 4.80$$

5. Convert 4.80 to feet and inches: multiply .80 of a foot by 12″ for an answer of 9.6″.

6. Convert .6″ to a fraction: multiply .6 by 16 for an answer of 10/16 inches, reduced to 5/8″.

ANSWER

$$4' - 9\frac{5''}{8}$$

Fig. 6-12. An example in computing the line or theoretical length of a rafter using the Pythagorean theorem.

practical application, this is not a great problem because the accepted margin of error in roof framing is 1/8 of an inch. I have also found that in the process of roof framing it is wise to use two different methods to calculate rafter length. Using one method as a check against the other helps reduce costly mistakes.

Common Rafter Ridge Cut

As in Fig. 6-13, the line length of the common rafter is defined

Fig. 6-13. When you use a ridge board, deduct for the thickness of the ridge.

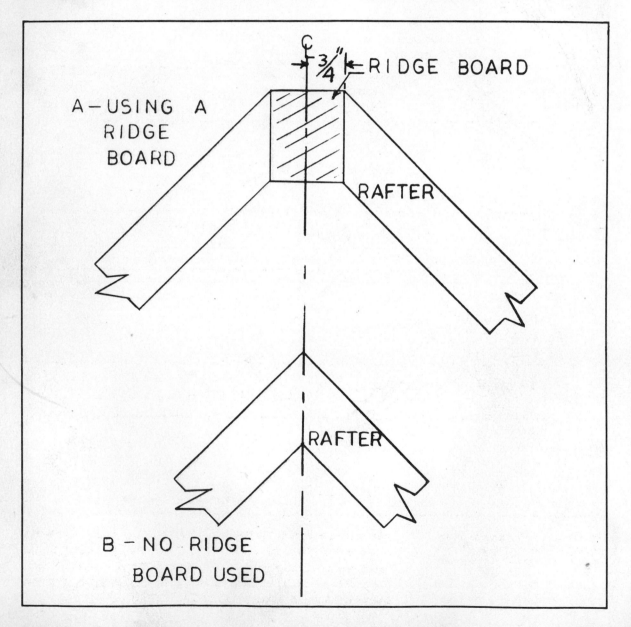

A—USING A RIDGE BOARD

RIDGE BOARD

RAFTER

B—NO RIDGE BOARD USED

RAFTER

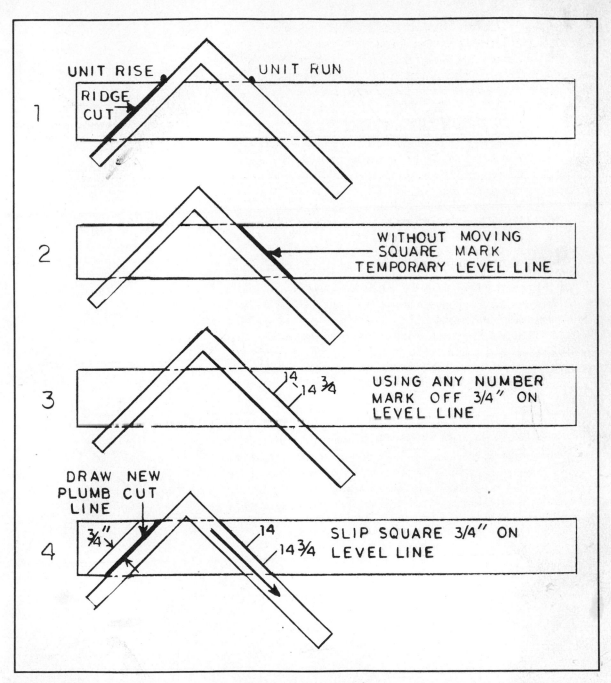

UNIT RISE   UNIT RUN

1   RIDGE CUT

2   WITHOUT MOVING SQUARE MARK TEMPORARY LEVEL LINE

3   14 3/4
    14 3/4
    USING ANY NUMBER MARK OFF 3/4" ON LEVEL LINE

4   DRAW NEW PLUMB CUT LINE
    3/4"
    14
    14 3/4
    SLIP SQUARE 3/4" ON LEVEL LINE

Fig. 6-14. Slip the square when allowing for the ridge board.

as a measurement from the outside edge of the wall plates to the center line of the ridge. A roof can be framed without using a ridge board, with the ridge ends of opposite rafters butting against each other, or they can be framed with the ridge ends of opposite rafters shortened and fastened to the ridge board.

1. If a ridge board is not used, no adjustment to the line length of the rafter has to be made.

2. When using a ridge board, one-half the thickness of the ridge stock has to be deducted from the rafter ridge end. This is known as shortening the rafter or deducting for the ridge allowance.

Slipping the Square

Figure 6-14 shows how to slip the square to make any deduction in a horizontal or level line. This measurement is always made at right angles (90 degrees) to the plumb line. In this example, it is assumed that the ridge thickness is 1 1/2 inches, which is the actual dimension of any lumber classified as 2-inch stock.

1. Place the tongue of the square on the plumb line, using the cut of the roof.

2. Draw a level line along the outside edge of the body of the square.

3. Without moving the square, mark on the board any number on the level line. Also place a mark 3/4 of an inch from the first mark, using the numbers 14 and 14 3/4 inches.

4. Slip the square carefully along the level line so that the 14-inch graduation is on the 14 3/4 mark on the rafter. Along the tongue of the rafter draw the ridge cut on the rafter. This move places the new ridge cut exactly 3/4 of an inch at right angles to the original theoretical ridge mark.

Rule for Horizontal Measurements on a Rafter

One of the fundamental rules for measuring in a horizontal direction on a rafter is that it must be made on a level line. When done in this fashion, the measurement will be always be the same for

Fig. 6-15. A rafter in a level line with zero angle.

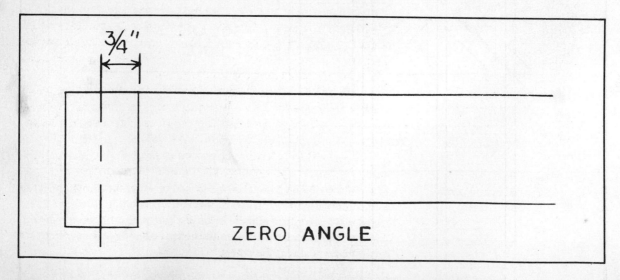

ZERO ANGLE

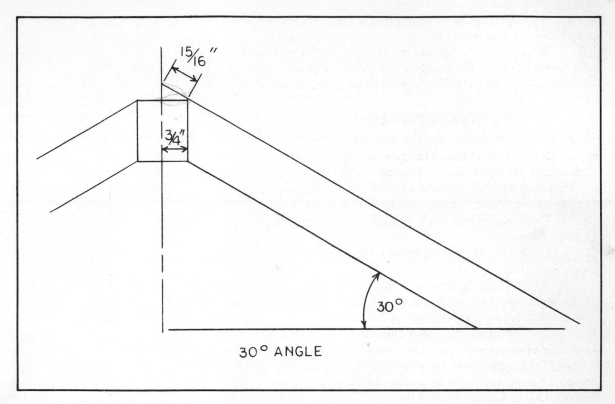

15/16"

3/4"

30°

30° ANGLE

Fig. 6-16. The difference in ridge deduction for a 30-degree angle.

any roof angle. Any measurement taken along slope line (hypotenuse) varies with each change of angle. All examples use a 1 1/2-inch-thick ridge, with one-half the thickness being 3/4 of an inch. This principle is shown in Figs. 6-15 through 6-18.

Figure 6-15 shows a rafter in a level line with zero angle. The deduction for a ridge thickness of 1 1/2 will always be 3/4 of an inch (which is one-half the thickness of the ridge).

Figure 6-16 shows the difference in ridge deduction measurement for a roof with a 30-degree angle in reference to a horizontal line. Measured on a level line, it is 3/4 of an inch; measured along the slope line (hypotenuse) it is 15/16 of an inch.

Figure 6-17 shows the difference in ridge deduction measurement for a roof with a 45-degree angle in reference to a horizontal line. Measured on a level line it is 3/4 of an inch; measured along the slope line (hypotenuse) it is 1 1/8 inches.

Figure 6-18 shows the difference in ridge deduction measurement for a roof with a 60-degree angle in reference to a horizontal line. Measured on a level line it is 3/4 of an inch; measured along the slope line (hypotenuse) it is 1 1/2 inches.

Notice how the measurement along the slope line increases as the angle increases while the measurement at right angles to the plumb line remains the same.

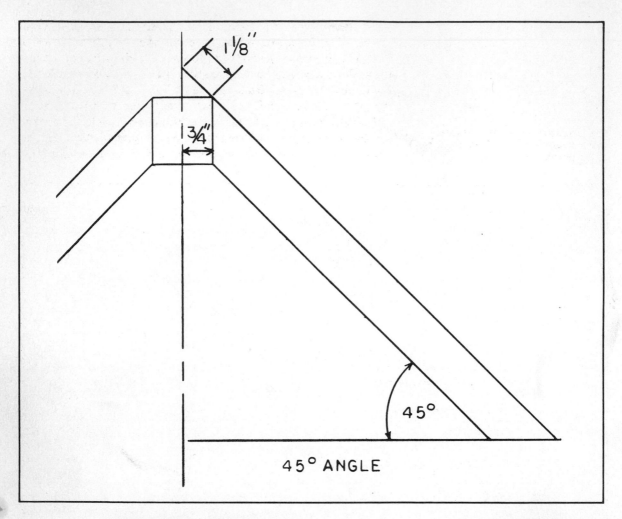

1 ⅛"

3/4"

45°

45° ANGLE

Laying Out the Bird's-mouth

Fig. 6-17. The difference in ridge deduction for a 45-degree angle.

Figure 6-19 illustrates two positions of a rafter in place on the top plates of a wall. The rafter labeled B in Fig. 6-19 is placed in position without any modification. Notice how the bottom edge of the rafter rests on the outside edge of the top plate. This is a poor method of framing because in time both edges will compress under the roof load, causing the rafter to drop creating a sag in the roof. The rafter labeled A in Fig. 6-19 has been modified with a bird's-mouth cut into the rafter. This creates a flat bearing surface to spread the load of the roof over a larger area and greatly reduces the possibility of the lumber compressing any sag in the roof.

To lay out the bird's-mouth (Fig. 6-20) using the cut of the roof:

a. Draw the plumb line at the ridge.

b. Measure the line length of the rafter, and mark it on the rafter.

c. Draw the plumb line at the measured line length (heel cut).

d. With the square set to the cut of the roof, draw the level line (seat cut) along the body of the square to the width of the top plate. For a 2 × 4 top plate at right angles to the heel plumb cut, it will be 3 1/2 inches from the plumb line to the bottom edge of the rafter. For a 2 × 6, the measurement will be 5 1/2 inches.

This completes the layout of the bird's-mouth. The dimension for the seat cut given above is the desired amount, but it might have to be varied if not enough stock remains between the heel

Fig. 6-18. All horizontal measurements must be made at 90 degrees to the plumb line.

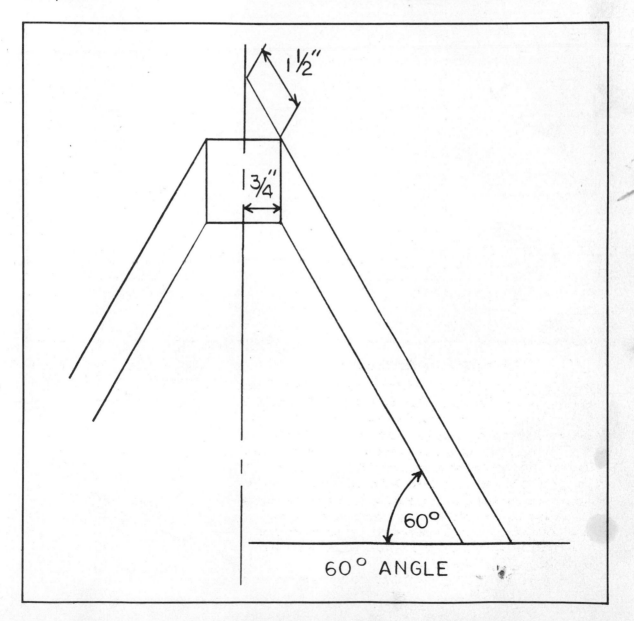

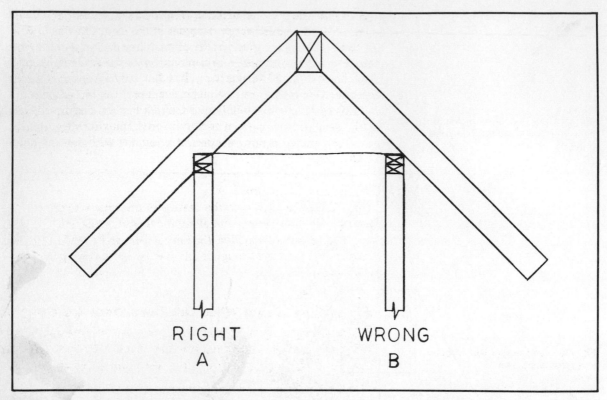

RIGHT
A

WRONG
B

of the bird's-mouth and the top edge of the rafter to support the weight of the overhang (E). At least one-half of the rafter depth should remain for the required strength. If possible, try to keep the seat (level) cut the same size as the top plate dimension. This will increase the strength of the rafter at that point.

Fig. 6-19. How the bird's-mouth spreads the concentrated load of roof rafter.

Overhang for a Common Rafter

The overhang is that part of the rafter that extends (projects) past

Fig. 6-20. Laying out the bird's-mouth of a common rafter.

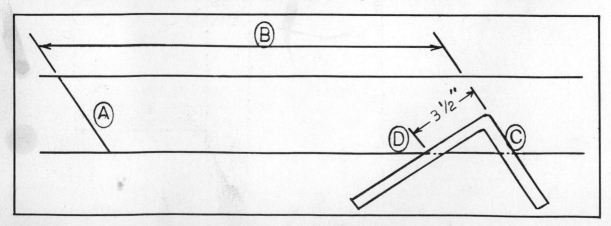

the outside walls of the building (Fig. 6-21). The overhang is determined by the architect or designer of the roof. The dimensions for the overhang are given in the plans. They can be shown in the plan view or shown on elevation or section views. When the dimensions given is stated as overhang, it is then only necessary to mark that given dimension on the slope line along the top edge of the rafter from the seat plumb line to the tail. Place the square on the mark at the slope of the roof and draw the tail line along the tongue.

When shown as a projection, it is laid off with the square as illustrated in Fig. 6-22.

a. Measure out the required dimension at right angles to the heel plumb line and mark it.

b. Move the square to the mark, set the square to the cut of the roof, and draw the tail cut along the tongue. Depending on the cornice style and construction, the tail cut can now be made. Usually it is a plumb cut even though it can just be cut square, giving the fascia a slope.

Laying Out a Rafter When Odd Dimensions Are Given

Not always do the given dimensions come out in even numbers. They frequently end in feet and fractions of a foot (inches). Figure 6-23 shows how to lay out a rafter when this problem occurs.

Fig. 6-21. The proper use of overhang and projection.

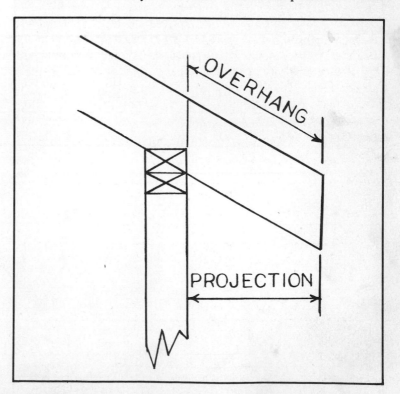

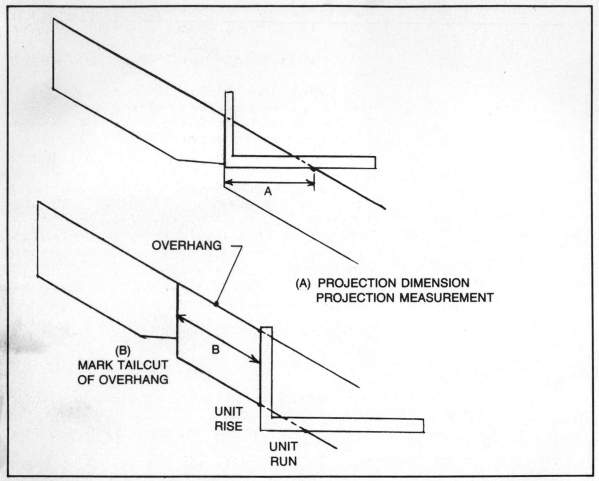

Fig. 6-22. How to measure and layout the tailpiece.

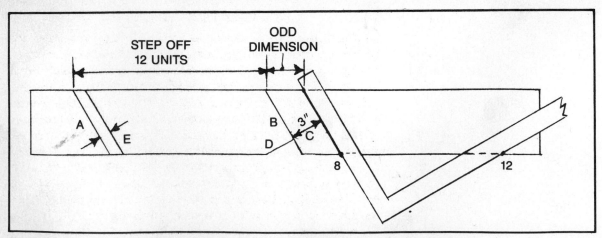

Fig. 6-23. Laying out of a common rafter with an odd run measurement.

Assuming that a roof is to be built that has a run of 12' 3" and a slope of 8 and 12, the steps in laying out the rafter are explained below.

a. Draw ridge cut line (A) using the square set to the slope of roof.

b. Step off 12 units. Mark and draw the heel plumb line (B).

c. Without moving the square measure, mark off the odd inches (3). Keeping the square set on the slope of 8 and 12, slip the square to this mark and draw the heel plumb line (D). This procedure produces a rafter with the correct line length for a rafter with a run of 12' 3".

d. Now lay out the bird's-mouth, lay out and mark the tail cut, and then make the deduction and mark the ridge cut. The layout of the rafter for an odd run is complete and can be cut.

At times it may be necessary to calculate the length of an odd dimensioned rafter mathematically. Following the steps listed below makes this operation quite simple.

a. On line one of the rafter table, find the per foot run for a rafter with a unit rise of 8 inches (14.42).

b. Convert 3 inches to a decimal of a foot, or a fractional equivalent by dividing 3 inches by 12, for an answer of .33 or 1/3 of a foot.

c. Multiply 14.42 by .33 or 1/3 for an answer of 4.76 (4 7/8) inches.

d. Subtract 4.76 from 14.42 for an answer of 9.66 (9 5/8) inches, the length of the overhang.

e. Calculate the line length of the rafter by multiplying 12 feet × 14.42, which equals 173.04 (14' 5 1/16").

f. Add 173.04 inches and the overhang of 9.66 inches for an answer of 182.7 inches (15' 2 11/16") for a total rafter length.

Laying Out a Rafter

Now that the fundamentals of calculating, laying out a rafter, and proper procedure for manipulating the square have been mastered, it is time to actually apply these skills, to lay out a complete rafter and cut it. For the first rafter, which will be a pattern for all other rafters, choose a board that is straight and has the least amount of defects. Figure 6-24 shows a comparison of a straight board and one that has a crown in it. You will find that the amount of crown does vary from one that is relatively straight to those boards that have a very pronounced curve to them. As the board is lying flat on a pile of lumber or across two saw horses, lift one and and sight down both edges. The crowned edge can than be determined. Mark the edge and continue the crowning process until all rafter stock has been checked. All crowns are placed up when the rafters are in position. If this is not done, the roof will develop a rippled effect

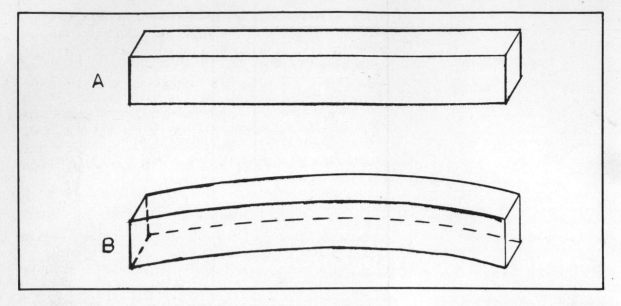

A

B

(plus create other problems). With the crown in the up position, the bird's-mouth will be cut on the opposite or lower edge of the rafter.

As with many operations in the construction of a building, there is more than one method of doing a task, and one method is no more correct than the other. Each craftsman has his favorite way of doing a particular job, and he will use this method because he is more comfortable with it. As you become more experienced, your favorite method will help you do the specific job with a minimum amount of effort and with the greatest amount of accuracy.

Some craftsmen like to work the crown toward them; others like the crown away from them. Some like to start at the left end of the board and work toward the right; others start at the right and work towards the left. Some like to start with laying out the overhang and work toward the ridge; others start at the ridge and work toward the overhang.

I am comfortable starting with the crown away from me, beginning the layout at the ridge working toward the overhang from the left end of the board to the right. This method allows you to visualize the rafter in its actual position, leaving any extra stock at the tail so that adjustments, if necessary, can be made to the overhang. Depending on existing conditions, some craftsmen cut the complete rafter on the ground. Others cut the ridge and bird's-mouth, leaving the tails run free, and then mark and cut the tails when the rafters are in place. Using this method, you are assured that there will be no variation in the projection and the fascia line will be straight and true. More tips and techniques are described in chapters that follow.

Fig. 6-24. The difference between a straight rafter and a rafter with a crown (exaggerated).

Examples of Laying Out a Common Rafter

Using the dimensions given in Fig. 6-25, lay out a common rafter. Given is a span of 26' 0", a slope of 10 and 12, and a a projection of 1' 4".

a. From table 1 on the face of the square, find the unit length (per foot run 15.62).

b. Divide the span in half to obtain the run (13 feet).

c. Multiply 15.62 by 13 to obtain the line length (203.06 inches).

d. Divide by 12 to convert to feet inches and fractions of an inch. Note the answer 16' 11 1/16" is the theoretical length or the measured length of the rafter.

The Actual Layout

Place the rafter stock, across two saw horses, with the crown away from you. Then follow the steps shown in Fig. 6-26.

a. Place the square on the rafter stock, with the unit rise of 10 on the tongue and the unit run 12 on the body, and draw the ridge cut along the tongue. Mark the unit run on the body.

b. Measure and mark off the line length of the rafter 16' 11 1/16" and draw the plumb line along the tongue of the square.

c. Flip the square over so that the tongue is on the plumb line

Fig. 6-25. A simple section view supplying the necessary information to lay out a common rafter.

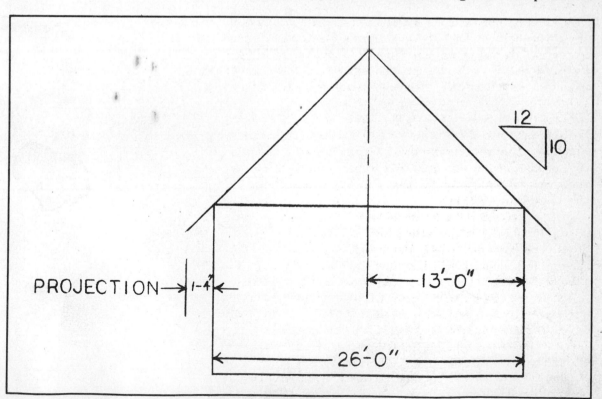

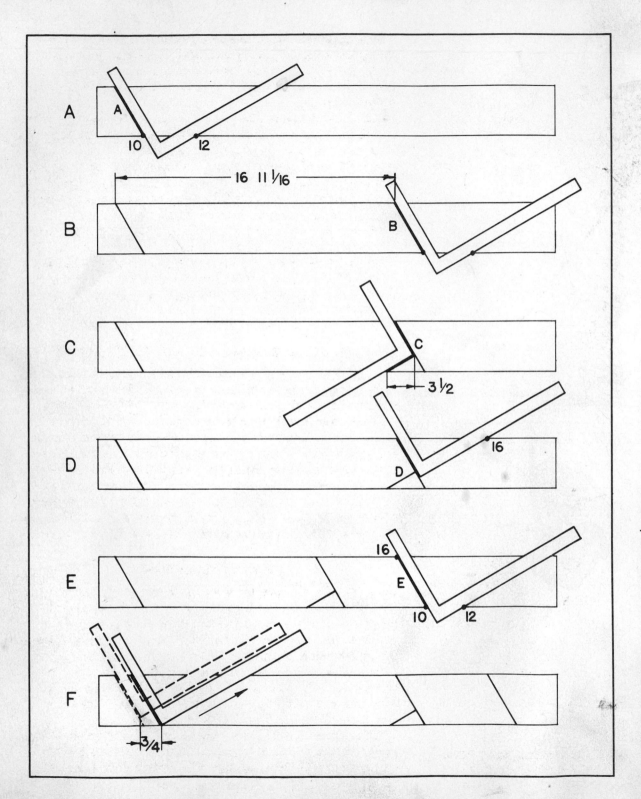

(that was just drawn) with the body pointing to the left. This places the square at right angles to the plumb line. Measure along the body of the square 3 1/2 inches from the plumb line to the lower edge of the rafter. Draw the seat cut of the bird's-mouth on this body of the square.

d. Flip the square over so that the body now points back to the right. With the tongue edge on the plumb line, measure 16 inches the length of the overhang and mark it.

e. At the 16-inch mark, set the square to the cut of the roof and draw the plumb line for the tail cut. *Note*: Steps (D) and (E) can be omitted if you want to first install the rafters and then mark and cut the tails when the rafters are in place.

f. Return the square to the ridge end. Place the tongue parallel on the plumb line. Slip the square to the right one-half the thickness of the ridge stock. In this example, it is 2-inch stock, with an actual dimension of 1 1/2 inches, divided in half for a dimension of 3/4 of an inch. Draw a new ridge plumb line that will be the new cutting line.

PUTTING THE ROOF PARTS TOGETHER

Now that the theory of roof framing and the principles involved in laying out the roof framing members have been mastered it is time to construct the roof. Check the calculations and layout lines. It is important that the first rafter be correct because this will be used as the pattern for the remaining rafters. If everything is correct, cut the rafter as marked (ridge line, bird's-mouth, and—if you chose at this time—the overhang). You could wait until the rafters are in place on the top plates of the building walls to align the tails and cut the overhang. I suggest that only the second rafter be traced from the pattern and cut. Later in this chapter, a method to check the accuracy of the rafter layout using these two rafters is detailed.

Making a Story Pole

On many jobs, a carpenter will use a device called a story pole. This is a jig used to help in the layout of various studs, joists and rafters. The story pole is a job-built device usually made using a single clean and straight 1 × 2 or 1 × 3. A single thickness is sufficient, but if it is necessary to obtain a greater length the pole can be made from two or more boards spliced as shown in Fig. 6-27. On this board, the position of each rafter is marked, the marks are then transferred to the outside wall plates, the ridge and, if necessary, the girder or weight-bearing interior partition.

Measure and cut the story pole to the length of the building. Divide this length in half to find the center of the building. Mark the center line on the story pole (Fig. 6-28) and then proceed to

Fig. 6-26. The actual steps involving in laying out a common rafter.

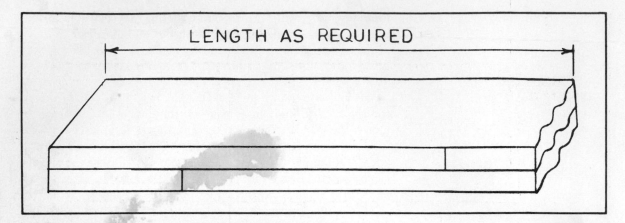

LENGTH AS REQUIRED

lay out the rafter positions 16 inches on center.

Start from the left end of the story pole and measure in 15 1/4 inches. Using a try square or combination square, draw a line across the board and place an × to the right of the line. This is the position of the rafter. From the line, measure and mark every 16 inches until the end of the board is reached. It doesn't make any difference if the last space is less than 16 inches. This is a common occurrence because rarely are the dimensions such that all measurements come out even.

Transfer the marks from the story pole to the top plates and the ridge. If necessary, use the center-line marks to align the story pole with the center line of the plates and the ridge (Fig. 6-29). If all the layout is done from left to right, that is floor joists, wall studs, ceiling joists and roof rafters on both front and back walls, this will place the second roof rafter against the second ceiling joist and all of the following ceiling joists and rafters should have the same pattern. All the framing member loads falling under each other, starting with the rafters and continuing down to the sill plate. This is the most desirable method of framing as it distributes the load directly through each framing member directly to the foundation. Unfortunately this method cannot always be used.

With the use of double top plates, the load of the upper struc-

Fig. 6-27. How to stager joints when a long story pole is needed.

Fig. 6-28. A simple story-pole layout for rafters 16″ on center.

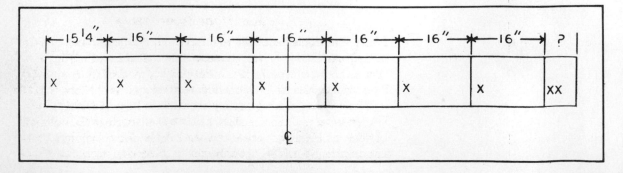

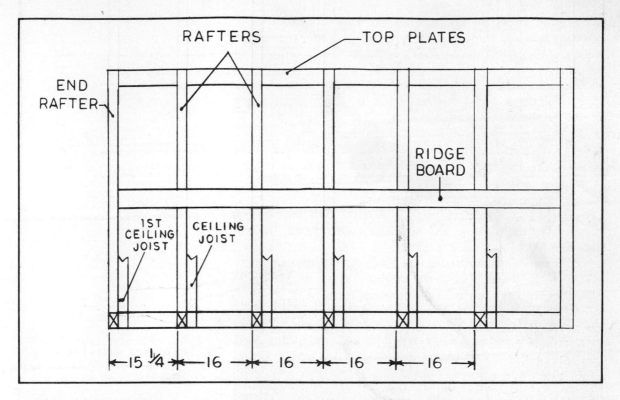

RAFTERS — TOP PLATES

END
RAFTER

RIDGE
BOARD

1ST
CEILING
JOIST

CEILING
JOIST

←15 ¼→ ←16→ ← 16 → ←16→ ← 16 →

Fig. 6-29. The actual placement of roof rafters in relationship to the ceiling joist when using 16″ center spacing.

ture is transmitted to the members below. Either way is an accepted standard in the construction industry. The first and last rafter (end rafters) are positioned flush with the outside edges of the end wall plates with a block next to them and then the ceiling joists. This places all ceiling joists and roof rafters in the proper place for the application of the Sheetrock on the interior and roof sheathing on the exterior. The use of the story pole might seem a bit cumbersome for rafter layout, but it will help in maintaining accuracy. Later, as you become accustomed to working on the top plates of the walls, you might choose to eliminate the use of the story pole. For the inexperienced person or if there are a number of the same layouts to be done, the story pole is a very handy device.

Erection of the Roof Frame

Before starting to install the roof rafters, it is best to check the two rafters that have been cut to make sure that the pattern layout is correct. To accomplish this, a flat surface is needed. It must be at least the width of the span and deep enough so that the rafters will fit on the surface when they are in place. I have found the most practical way to do this is place a number of sheets of the plywood that are to be used for the roof sheathing on the ceiling joists and tack them in place. This not only creates a working surface on which to check the rafters, but it can also be used as a platform from which the carpenter can work in safety to place the rafters. After the rafters are installed, the plywood can be passed up and used as sheathing.

Caution: Make sure the ends of the plywood rest on a joist so that the platform is solid, and remember that tack in place means to nail temporarily. A 8d nail in each corner will do. Figure 6-30 illustrates the proper procedure to check the rafters.

a. Snap a chalk line on the deck longer than the span of the roof. On this line mark off the dimension of the span.

b. Tack two pieces of wall plate material in place so that the outside edges are on the span marks.

c. Nail a small piece of the ridge material on to the ridge end of one rafter and place the two rafters in position so that the seat cut rests on the wall-plate material.

d. If the layout and cutting was done properly, the two rafters will be in the proper position and the joints at the ridge and wall plates should fit tightly (no spaces around them).

e. The final check will be to make sure that the ridge is at the correct height. To accomplish this, multiply the unit rise in inches by the run in feet. The answer will be in inches. To this, add the amount of stock that remains on the plumb line from the top edge of the rafter to the level cut of the seat.

Fig. 6-30. Two rafters being checked for proper fit, using a temporary platform and pieces of ridge and plate stock to complete the jig.

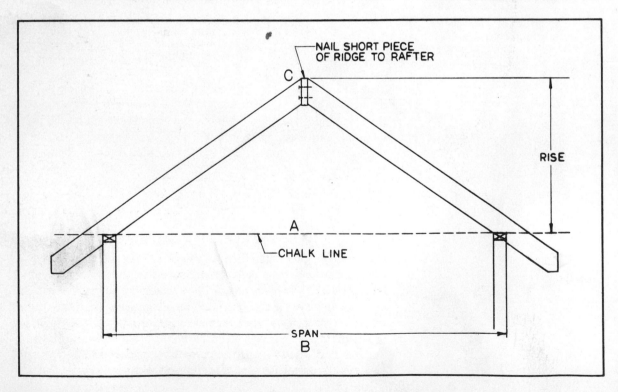

Fig. 6-31. Common rafters. Notice how the saw cut penetrates the rafter stock. This is not the best practice to use in cutting a rafter, but it is common and accepted in construction today.

f. Find the correct dimension of the rise of the roof. Measure this distance from the chalk line to the top edge of the ridge. If everything is satisfactory, use the rafter pattern and cut the remaining rafters. If it is not satisfactory, check the layout and make the needed corrections.

Figure 6-31 illustrates a common rafter with the bird's-mouth cut ready to go on the building. Notice the way the bird's-mouth is cut. On one rafter the cut was made with only a power saw. Observe how the saw kerf goes past the plumb and heel cut into the rafter stock. This is normal when using a circular saw. On the other rafter, the power saw was stopped before the blade reached the seat cut line, and the cut was finished with the hand saw. In practical application, both methods are used. The use of the power saw only is an accepted method (though not the most desirable). It is used mostly on project housing. On good, custom workmanship, the use of the two saws is customary.

Figure 6-32 shows rafters stacked ready to be put in place. A few points to remember are that the ridge cut goes up so that as the rafter is pulled up the ridge end is in position to be nailed. Never leave rafters in this position for a long time. Stacking them in this fashion for a few days will increase the possibility that the rafters will warp.

The tools needed to install the rafters are the framing hammer, cats-paw, level, and plumb bob. Nails needed are 8d, 10d, and 16d common. You will also need a number of pieces of material for bracing. It can be 1 × 3 or 2 × 4 stock that is used only temporarily and can be used later on the job. It is suggested that the crew consist of four men, but the job can be completed with as few as two men.

Figure 6-33 shows a temporary deck in place under the ridge to allow the carpenters to work in comfort with relative safety.

Notice the bracing used to keep the rafters plumb. They will remain in place until the roof is sheathed. If the top edge of the ridge is more than 5 feet above the working deck, a temporary scaffold platform should be built. The easiest way to make this platform is by placing sturdy planks across sawhorses. For quick construction, patented sawhorse brackets can be purchased in most building supply houses, lumberyards, or hardware stores. Use of these brackets simplifies and speeds up the construction of sawhorses.

Fig. 6-32. Rafters stacked ready to be placed. Notice the framing for a window in place to create a dormer.

Fig. 6-33. The ridge board and some rafters in place. Notice the temporary platform and bracing being used.

There are two methods used to place the ridge and rafters. The choice of methods depends on existing conditions and preference of the person doing the work. Regardless of which method is chosen, it is suggested that the actual height of the top edge of the ridge first be determined and a few temporary upright supports be prepared.

The theoretical framing point of the total rise is approximately the middle of the ridge board. To determine the actual height of the top edge of the ridge, the distance between the top of the plate (seat cut) measured on the heel plumb line must be added to the total rise. Fig. 6-34 shows an example of this problem.

a. Assuming a total rise of 53 inches.

b. The measured distance of 3 inches from the top of wall plates to the top edge of the ridge on the heel plumb line.

c. Adding 53 inches plus 3 inches gives a total of 56 inches, the actual height of the top edge of the ridge above the wall plate.

Figure 6-35 illustrates a temporary support to hold the ridge in both vertical and longitudinal directions. This support can be used with most any method of ridge placement.

Figure 6-36 illustrates the method used to fasten the rafters to the ridge board. Place ridge board on the temporary deck, bring the rafters up to the ridge, and face nail through the ridge. The rafter can be nailed at each end or one rafter in from the end as shown. It will seem awkward at first because of the slant of the ridge, but if one end is done at the time it can be accomplished.

Fig. 6-34. How to calculate the actual height of the ridge board above the top plates as related to the framing triangle is shown by the dashed lines.

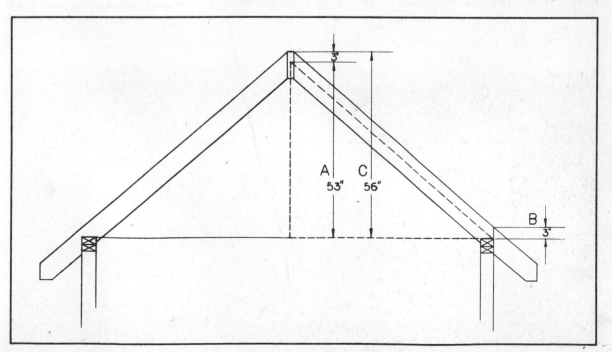

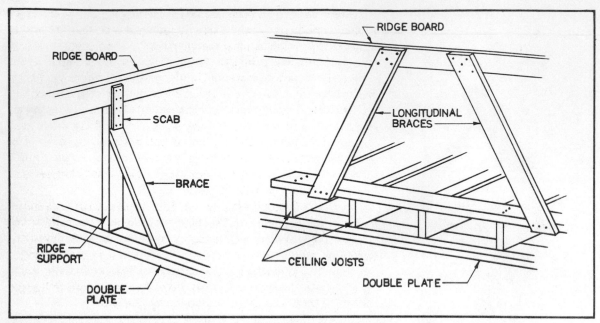

Fig. 6-35. Methods of supporting and bracing a ridge board prior to placing the rafters.

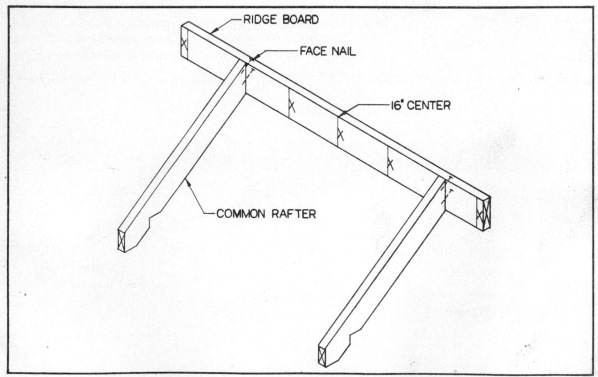

Fig. 6-36. Two rafters nailed to the ridge when using the alternate method of placing ridge board and rafters in place at the same time.

Place the upright supports under the ridge until the tail ends of the two rafters are nailed into place. Make sure that the bird's-mouth is seated tight against the top plate. The two opposite rafters can now be toenailed in place at the ridge and bird's-mouth. The ridge and rafters will now stand in place without collapsing. Watch that it does not move in a longitudinal direction (which it can do) until braced. Hang a plumb line on one end of the ridge and install longitudinal braces when the point of the plumb bob is exactly over the outside edge of the top plate (Fig. 6-37).

If all calculations are correct and care is used in cutting and erecting the rafters, the ridge should be level when checked and all rafters should be plumb.

The second method of erecting the roof ridge board, is to place the ridge in position first (Fig. 6-38). It is kept in position by the upright supports that are braced in longitudinal and transverse directions. It is important to check and make sure that the ridge is centered on the span and that it is level. Notice the splice block used to connect two pieces of ridge stock when one single full length is not available to run the full length of the roof.

The splice block is best made of 1/2-inch plywood, with one block on each side of the ridge. The top edge of the splice block is kept about 3/4 of an inch below the top edge of the ridge. Before butting together and nailing, check that the ends of the ridge stock are cut square and the top edges of both ridge pieces are in a level line. With the ridge in place, centered level and braced, the

Fig. 6-37. A ridge board that has been spliced and put in place together with the rafters. Notice the bracing used.

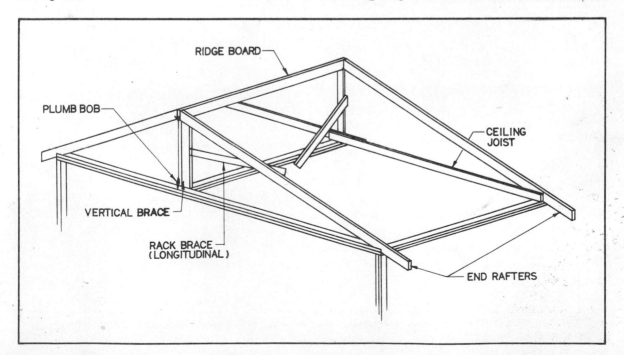

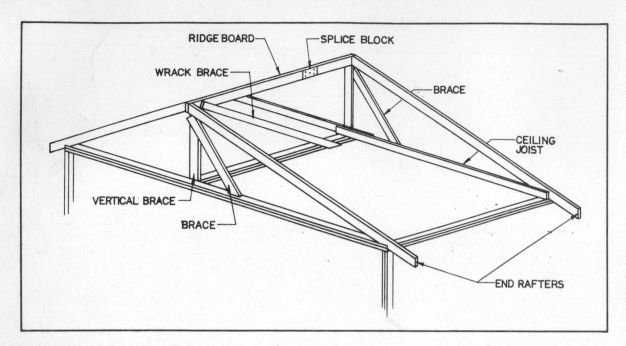

RIDGE BOARD

SPLICE BLOCK

WRACK BRACE

BRACE

CEILING JOIST

VERTICAL BRACE

BRACE

END RAFTERS

rafters can now be nailed in place. Nail the tail end first and then the ridge end.

It is important to place rafters on alternate sides of the ridge. If rafters are placed only on one side of the ridge, the weight might have a tendency to push the ridge out of place or even cause it to collapse.

Figure 6-39 shows the ridge and rafters in place on the roof. Notice the longitudinal brace in place on the right-hand side and a few of the gable studs in place on the left side.

Fig. 6-38. A ridge board and four rafters in place being checked for plumb with the use of a plumb bob. Notice that only longitudinal bracing is necessary.

Fig. 6-39. Roof rafters in place and braced. Tails are trimmed and some gable studs are in place.

Collar Beam

The collar beam, also called a collar tie, is a roof-framing member that is fastened in a horizontal position between two opposite roof rafters (Fig. 6-40). The stock can be 1 × 4, 1 × 6 or 2 × 6 and is usually fastened about one-third of the way down from the ridge. Because this distance down is not a specific dimension, the collar beam is sometimes placed so that it can act as a ceiling joist in a finished attic. At other times it will be desirable to eliminate the collar beam completely. In order to meet building codes, it then becomes necessary to use a steel strap on the top edges of the rafters and ridge board. The ends of the collar beams are cut to the slope of the roof and fastened in place so that they are even with or slightly below the top edge of the rafter.

The collar beams act as an extra brace for the rafters and help keep the walls of the building from being pushed out by the weight of the roof. Because sizes and positions of these members are not crucial their dimensions can be determined by placing them in an approximate position and making a pattern. The collar beam can also be calculated and laid out with the use of the square. By following the steps below and using the dimensions shown in Fig. 6-40, the need to climb up on the roof and make a pattern is eliminated. Climbing on the roof can become cumbersome at times, depending on the size of the collar beam and the height of the roof above ground.

Fig. 6-40. Proper placement of a collar beam. Also shown is the proper use of steel strapping, if required.

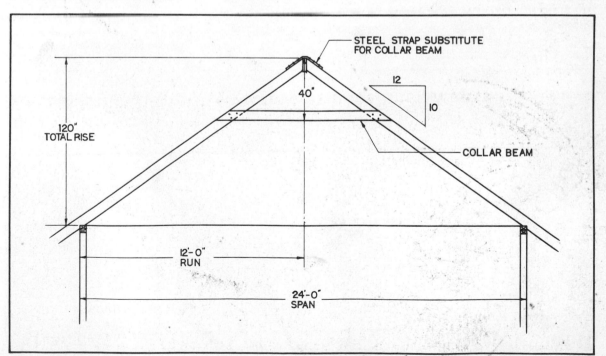

a. The distance down from the top plate to the top of the ridge was determined to be 40 inches by taking 1/3 of the total rise (120 inches). Notice that the measurement is taken from the top edge of the ridge to the bottom edge of the collar beam.

b. Multiply this distance (40 inches) by 2 and divide the result by the unit rise (10 inches) for a result of 8 feet.

c. To find the actual length, multiply 8 feet by 2 for an answer of 16 feet.

To get the correct angle cut on the ends of the collar beam, set the square on the end of the stock as shown in Fig. 6-41. Using the slope figures (10 and 12), mark along the body of the square. Measure off 16 feet from the long point of the cut and repeat this step. This will give you a collar beam cut to the proper length and angle.

Gable Studs

Gable studs are vertical members that are placed in the gable end of a house. They extend from the top plates to the underside of the rafters. The primary function of the gable stud is to supply a nailing surface for the wall sheathing. Gable studs add little if any structural value to a building, and they can be removed completely if required. Shown in Fig. 6-42 are gable studs in place as viewed from inside the house. Gable studs are placed either 16 or 24 inches on centers. Laying out and finding the length of the gable studs can be accomplished in a number of different ways. Figures 6-43 through 6-45 show three of the most popular layout patterns for gable studs.

Fig. 6-41. The proper way to use the square to mark angle cuts on the ends of collar beams.

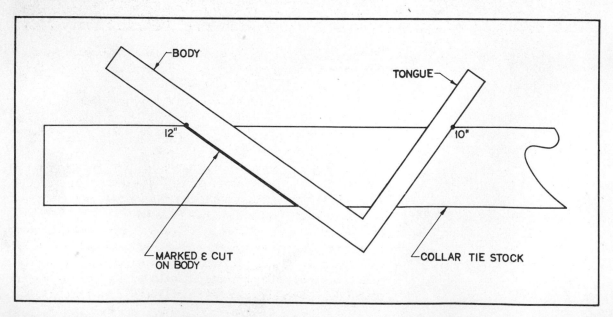

Fig. 6-42. Close-up view of gable studs in place.

Figure 6-43 shows the gable stud centered directly below the ridge board. The remaining studs 16 inches on center in both directions to the outside of the top plates.

Figure 6-44 shows the two longest studs placed 8 inches from the center line, in both directions, with the remaining studs 16 to the outside of the top plates. This method is used mostly when a rectangular louvered vent is to be used.

Figure 6-45 shows the gable studs directly over the studs in the framed wall, just below the gable. This method is not the most popular one because of the stud location. It might take a number of measurements to cut the studs properly.

The layout methods shown in Figs. 6-43 and 6-44 can be used by applying the common difference in length method formula for gable studs and adding this difference starting with the shortest stud or subtracing this difference when starting with the longest stud. All the remaining gable studs can then be cut at one time and

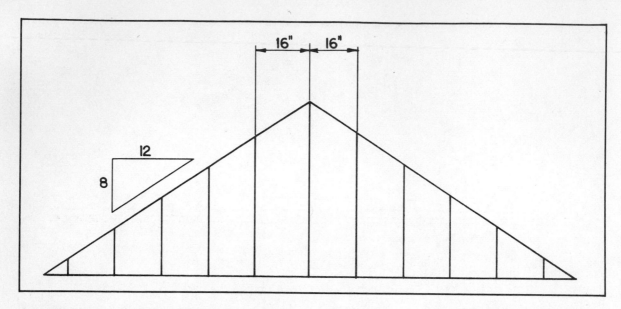

they will fit their respective places on the gable wall. More details on this method are described later in this chapter.

The easiest method to find the length of a gable stud is to physically measure it.

a. Determine the distance from the base of the stud to the underside of the rafter. Make sure the measurement is made with the stud in a plumb position.

b. On one edge of the stud, using the cut of the roof, place the square with the body to the top of the stud and mark a line along the body. Measure the length of the stud from the high point of the top end to the base and draw a square line at this point. The stud is then cut. The top cut is at an angle and the bottom cut is square. The stud is then toenailed to the rafter and plates.

Fig. 6-43. Gable studs laid out starting with the center one out to the outer walls.

Fig. 6-44. Gable studs starting with the longest ones 8 inches either side of the center of the house to the outer walls 16 inches on centers.

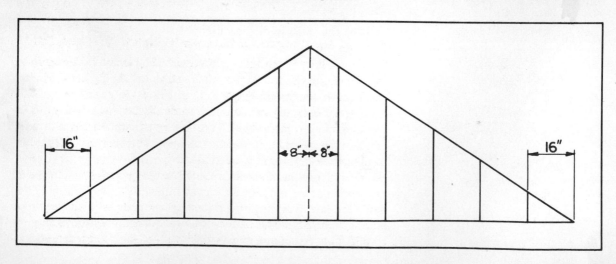

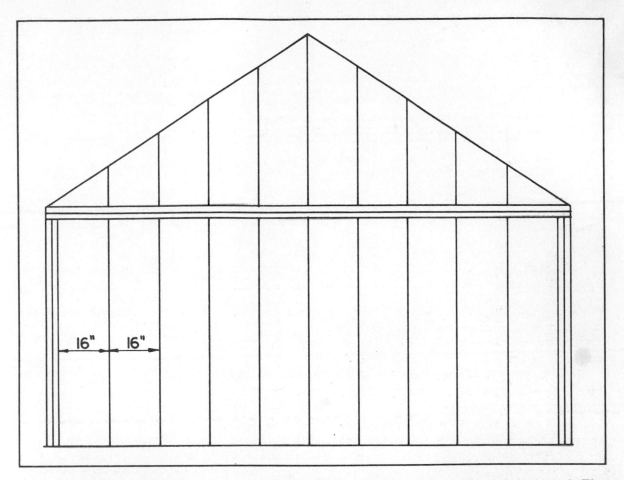

Fig. 6-45. Gable studs placed directly over wall studs below.

Another popular method of cutting gable studs is shown in Fig. 6-46. The gable stud is cut from the top plates to the top edge of the end rafter. It is then notched to fit around the rafter.

Finding the Common Difference in Length of Gable Studs

Figure 6-47 illustrates the principle of the common difference in length of gable studs. Triangles A, B, C and A, D, E are similar triangles. Therefore, the sides C, B, and D, E are proportional, making the difference in length of the sides the common difference in the length of the gable studs. This difference can be applied to each adjoining gable stud, eliminating the necessity of measuring each stud separately. Similarly placed studs in each section on both ends can be cut at the same time. This method helps move the job along quicker.

There are two methods of finding the difference in length of gable studs. The difference can be calculated mathematically or by slipping the square. Figure 6-48 shows how to slip the square

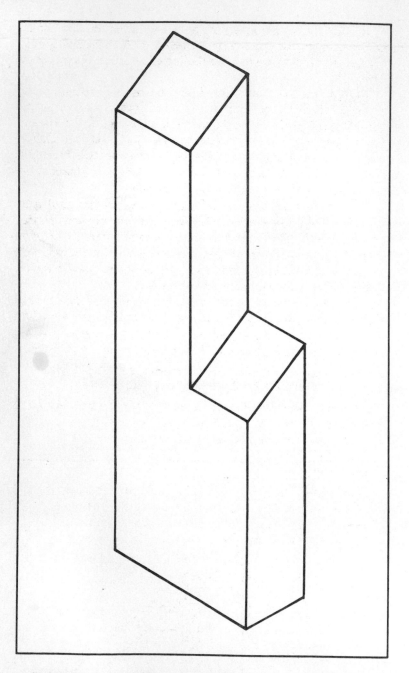

to find the common difference of gable studs for a roof with a cut of 6 inches.

    a. On the edge of a board, place the square to the cut of the roof, 6 inches on the tongue and 12 inches on the body.

    b. Mark a level line along the outer edge of the body. Also mark the unit run (12 inches) on the edge of the board.

Fig. 6-46. Detailed drawing of the top of a notched gable stud.

c. Slip the square along the level line in the direction of the arrow until the stud spacing (16 inches) reaches the position of the unit run mark. Be sure to keep the top edge of the square on the level line.

d. The square should now be in the position shown by the dashed lines. With the body set on the 16-inch mark, the tongue will automatically be at the 8-inch mark on the edge of the board. This measurement of 8 inches indicates the common difference of the gable studs. If the first stud is the longest one, this difference is subtracted for the next stud. If the first stud is the shortest, the difference is added for the next longest stud.

The formula for common difference states that the stud spacing (16 or 24 inches o.c.) multiplied by the unit rise and divided by the unit run (12 inches) equals the common difference. For the example shown in Fig. 6-48, a stud spacing of 16 inches o.c., with a unit rise of 6 inches, is used. Applying the formula stated above, it is found that the common difference is 8 inches.

Knowing the length of one gable stud, either the longest or the shortest, the common difference then can be applied to find the length of the remaining studs. All can be cut at one time. You will need four studs of each size to fill in all sections of the gable ends.

Framing an Opening in a Gable Wall

The same principle used to frame an opening in a wall is applied to framing an opening in the gable end of a roof.

a. Determine the position of the opening.

Fig. 6-47. The theory of the common difference in length of gable studs is obtained using the principle of similar triangles.

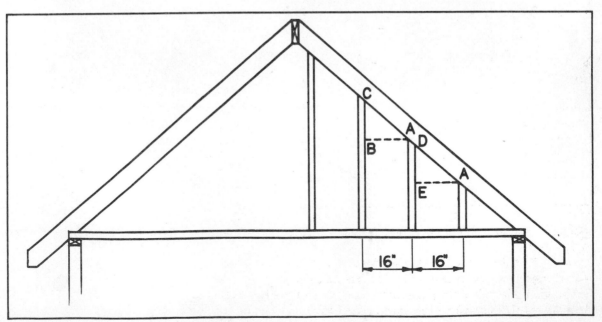

b. Place a gable stud on each side of the opening.

c. Block in the opening by placing a header and sill in position at the required height of the opening.

d. Fill in any cripple gable studs so that the required spacing is maintained.

Framing an Opening for a Triangular Louvered Vent

Figure 6-49 shows an opening framed for a triangular louvered vent in a gable wall.

a. Measure the base of the louver. Measure to the inside of the flange, the nailing flange will be fastened to the rafter and sill plate of opening.

b. Cut the ends of the sill plate to this dimension using the slope of the roof and install them.

c. Cut and install the required cripple gable studs and fasten in place under the sill plate. Both ends of these studs are cut square.

d. With the square set to the slope of the roof, cut the top angle of the full gable stud. Mark the required length and cut the bottom using a square cut. Applying the common difference, measure and cut the remaining studs and fasten them in place.

When placing the gable studs, make sure that the center spacing of the studs is maintained regardless of where the studs fall

Fig. 6-48. Finding the common difference in length of gable studs by slipping the square.

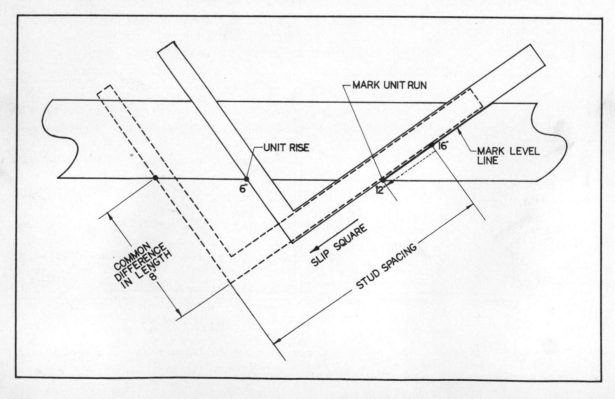

MARK UNIT RUN

UNIT RISE

16"

MARK LEVEL LINE

6"

12"

SLIP SQUARE

COMMON DIFFERENCE IN LENGTH 8"

STUD SPACING

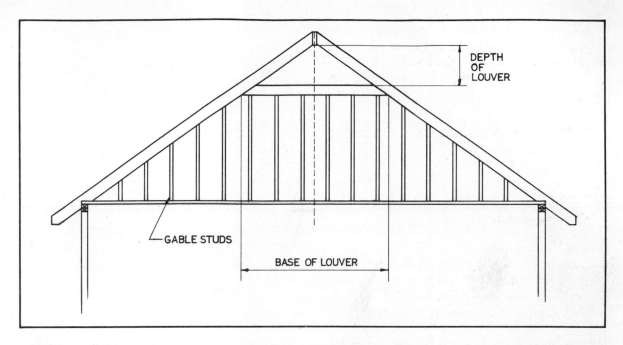

DEPTH OF LOUVER

GABLE STUDS

BASE OF LOUVER

Fig. 6-49. An opening for a triangular louvered vent framed in a gable wall.

for the openings (the same as is done when framing openings in a wall). See Figs. 6-49 and 6-50.

The framing for the roof is now complete. Continue the wall sheathing up the gable ends to close them in. Sheathing the roof is explained in the next chapter.

PROGRESS CHECK

1. A good scale for a roof framing plan is _____ or _____ equal to 1 foot.

2. To construct a roof, a carpenter must have the following three pieces of information _____, _____, _____.

3. The theoretical length of a rafter is also called the _____ or _____ length.

4. The tail piece is another name for overhang or projection T — F.

5. Name the three parts of a bird's-mouth _____, _____, _____.

6. Finding the line length of a rafter mathematically is more accurate than stepping off? T — F.

7. The unit rise is always placed on the _____ of the square and the run is placed on the _____.

8. To find the rise of a roof multiply the _____ by the _____.

9. A rafter must be shortened to allow for the _____ of the _____.

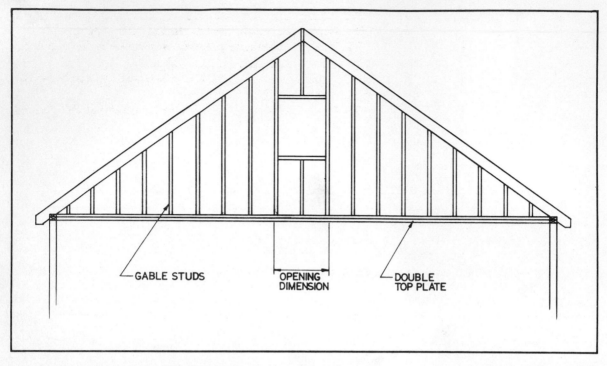

GABLE STUDS

OPENING
DIMENSION

DOUBLE
TOP PLATE

10. Measuring in a horizontal direction on a rafter is always done on a _____ line.

11. The seat cut of the bird's-mouth from the plumb line to the bottom edge of the rafter is usually 3 1/2 inches. T — F.

12. Any measurement can be used to lay out the overhang. T — F.

13. The projection of a rafter is measured at right angles (90 degrees) to the heel plumb line. T — F.

14. The dimensions for the overhang or projection are shown in the plans. T — F.

15. The crown of a rafter is always placed up, which is the opposite edge of the bird's-mouth. T — F.

16. Calculate the length of a common rafter with a run of 12' 4" and a cut of 8 inches, with a projection of 6 inches.

17. A story pole is absolutely necessary when laying out a roof. T — F.

18. Before cutting all the rafters, it is best to check the pattern rafter for proper fit. T — F.

19. It is not necessary to construct a temporary working platform when starting to erect the roof. T — F.

20. Rafter location must be accurately marked on wall plates and _____ _____ for a well-erected roof.

21. After the ridge board and four rafters are in place, it is wise to plumb, level, and _____ the roof.

Fig. 6-50. A rectangular opening framed in a gable wall.

22. A _____ _____ is used on every other rafter to give extra strength to the roof structure.

23. The collar beam helps the floor joists keep the walls from being _____ _____.

24. Gable studs are used in the end walls to help bear the weight of the roof. T — F.

25. Openings are usually framed in the gable end walls to hold _____.

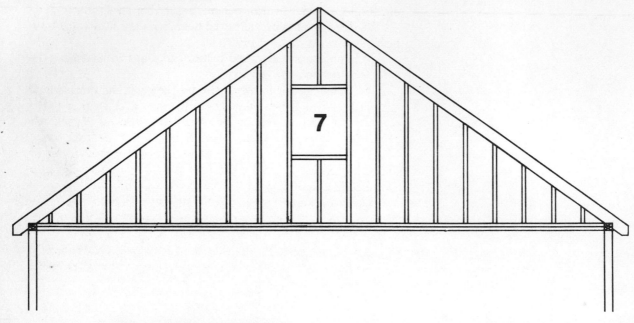

Roof Sheathing

Objective. Upon completion of this chapter, you should be fully familiar with the material called roof sheathing, the purpose of sheathing, the different kinds of materials used, and how to apply them. The following topics are described in this chapter:

- ☐ Lumber sheathing.
- ☐ Plywood sheathing.
- ☐ Decking or planking.
- ☐ How to lay out and install sheathing.
- ☐ Nailing patterns for sheathing.

Sheathing is applied over the skeleton of the roof formed by the rafters. It has a multipurpose function but it is mainly used to cover and protect the structure. Sheathing adds structural strength to the building by tying the roof members together, thus stiffening the building. Sheathing also supplies a nailing surface for the roofing material that is placed over it.

PLYWOOD SHEATHING

Plywood is one of the most common materials used for roof sheathing today. It is manufactured in 4- × -8 foot sheets in a variety of thicknesses, grades, and qualities. For sheathing work a lower grade called CDX is usually used. Large areas (32 square feet) can be applied at one time and its great strength relative to other sheathing materials makes it a most desirable choice for the job. At this time, you might want to review Chapter 1 on materials for a detailed description of plywood.

The thickness of plywood to be used for roof sheathing is determined by a number of factors:

1. The distance between rafters (spacing) is a most important factor. The larger the spacing the heavier the thickness of sheathing is used. When 16-inch o.c. rafter spacing is used, the minimum recommended thickness is 5/16 or 3/8 inches.

2. The type of roofing material to be applied over the sheathing plays a role. The heavier the roof covering the thicker the sheathing is required.

3. Prevailing weather conditions also determine sheathing thickness. In areas of heavy ice and snow loads, thicker sheathing is required.

4. Allowable dead and live roof loads are established by calculations and tests. These are the controlling factors in the choice of roof sheathing materials.

INSTALLATION OF PLYWOOD SHEATHING

Plywood sheathing is applied after the rafters, collar ties, gable studs and extra bracing, if necessary, are in place. At this time, it would be wise to make sure that there are no problems with the roof frame. Check rafters for plumb, make sure there are no badly deformed rafters, and check the tail cuts of all the rafters for alignment. The crowns on all the rafters should be in one direction: up.

Figure 7-1 illustrates two common methods of starting the application of sheathing at the eaves of the roof.

a. Here the sheathing is started flush with the tail cut of the rafters. Notice that when the fascia is placed the top edge of the fascia is even with the top of the sheathing.

b. The sheathing overlaps the tail end of the rafter by the thickness of the fascia material. Notice here the edge of the sheathing is flush with the fascia.

If you choose to use method *a* to start the sheathing, measure up on the two end rafters the width of the plywood panel (48 inches). From the rafter tail ends and using the chalk box, strike a line on the top edge of all the rafters. If method *b* is used, measure up the width of the panel minus the actual thickness of the fascia material. This chalk line will be used to position the upper edge of the sheathing panels. If the roof rafters are at right angles to the ridge and plates, the use of this line will place the sheathing panels parallel to the outer ends of the rafters.

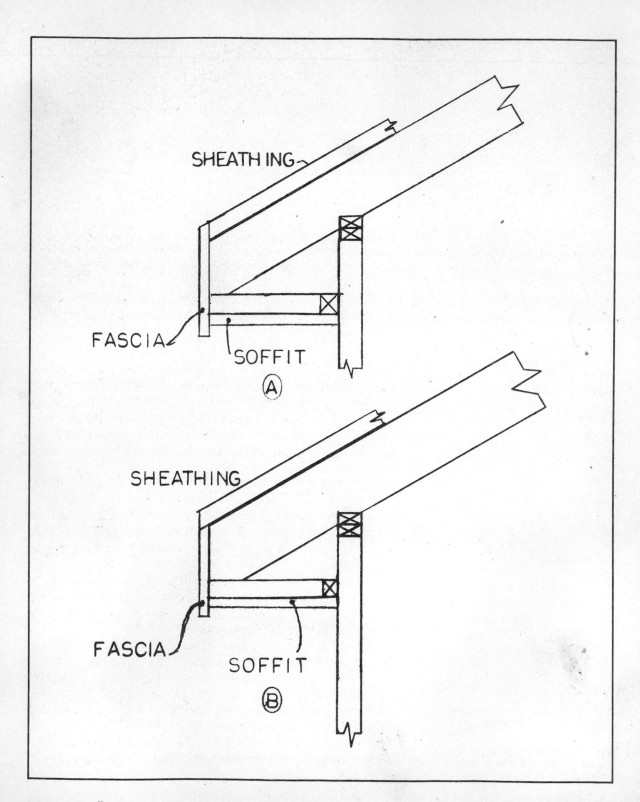

SHEATHING

FASCIA

SOFFIT

Ⓐ

SHEATHING

FASCIA

SOFFIT

Ⓑ

Fig. 7-2. The proper method of applying sheathing. Notice how the sheets are staggered.

Fig. 7-1. Left: Two methods of starting the first sheet of sheathing at the eaves of a roof.

Placing The Sheathing

Figure 7-2 shows how to start the plywood sheathing. Notice that this sheathing is being placed before the trim is applied. Sheathing is always placed from the lower (eaves) edge of the roof up toward the ridge. It can be started from the left side and worked toward the right, or you can start from the right and work toward the left. Usually it is started at the same end of the house from which the rafters were laid out.

☐ The first sheet of plywood is a full 4-×-8 panel. The top edge is placed on the chalk line.

☐ If the sheathing is started from the left side of the roof, make sure the right end falls in the middle of a rafter. This must be done so that the left end of the next sheet has a surface upon which it can bear weight and be nailed.

☐ The plywood is placed so that the grain of the top ply is at right angles (perpendicular) to the rafters. Placing the sheathing in this fashion spans a greater number of rafters, spreads the load, and increases the strength of the roof.

☐ The sheets that follow are butted to each other until the opposite end is reached. If there is any of the panel hanging over the edge, it is trimmed after the panel is fastened in place. A chalk line is snapped on the sheathing flush with the end of the house, and the panel is then cut with the circular saw. It is important that the manufactures specification stamp be read and proper spacing be allowed at the ends and edges of the sheathing. This is to compensate for any swelling that might take place with changes in moisture content.

☐ The cutoff piece of sheathing can be used to start the second course (row of sheathing) provided it spans more than two rafters. If it doesn't span two rafters, start the second course with a half sheet (4 × 4) of plywood.

☐ It is important to stagger all vertical joints. All horizontal joints will need blocking placed underneath or a metal clip (ply-clip) be used. The use of blocking and clips is determined by the rafter spacing and local building codes.

☐ This pattern is then carried to the ridge. The final course is fastened in place, a chalk line is snapped at the top edge of the rafters, and extra material cut off.

☐ The opposite side of the roof is then sheathed using the same pattern.

☐ The nailing pattern is as follows: the nails are spaced 6 inches o.c. along the perimeter of the sheet, the field is nailed 12 inches o.c. For plywood up to 1/2 inch 6d common coated, spiral, or ring nails are used. For plywood 1/2 to 1 inch thick, use 8d common coated, spiral, or ring nails.

Caution: It might become difficult to keep your footing on the roof, depending on the steepness of the angle, under these conditions. It is wise to nail lengths of 2-×-4 stock at intervals to use as steps and also act as a stop in case you do slip. When fastening these blocks, make sure nails go into a rafter. The blocks have a tendency to work loose if fastened to the thin sheathing only. It is also important to be careful when sawdust accumulates on the sheathing. When working with an electric saw, dust accumulates under foot and the deck does become slippery.

If the roofing material is not to be applied immediately after the roof is sheathed, cover the deck with building felt paper. In case of rain, the felt paper will prevent the sheathing from becoming wet. Otherwise, the plywood will delaminate and this causes the plies to separate. Figure 7-3 shows the effects of plywood sheathing having been exposed to rain.

Lumber Roof Sheathing

Before plywood became popular for roof sheathing, board sheathing was used for roof sheathing as well as for walls. Even today board sheathing is used for certain roofs such as wood shakes or wood shingles or possibly a slate roof.

If the roof is a pitched roof, 1-inch, nominal-sized boards are used. The width of the boards are usually 1 × 4, 1 × 6 or 1 × 8 laid at right angles to the rafters. The boards can be placed with edges butted without any space between them. This is called closed sheathing. They can be placed with a space between courses and this is called open sheathing.

Open sheathing is usually 1-×-4 boards with the on center spac-

Fig. 7-3. Plywood that has been exposed to rain will delaminate.

ing controlled by the weather exposure of the shingle. This type of sheathing is used where snow loads are not too great or in high-moisture areas. The sheathing is laid closed just past the overhang of the roof. The boards are random length with joints falling on rafters. Two 6 penny nails are used in each rafter for 1-×-4 and 1-×-6 boards. Joints should be staggered on rafters, with no two adjoining joints falling on the same rafter.

Closed sheathing is used where the roof covering requires a solid support underneath, or in areas where snow and icing conditions are severe. The boards are 1 × 8 inch and have three 6 penny nails on each rafter.

Use lumber that is well seasoned in order to minimize shrinkage and warping as the wood dries out. Fir, hemlock, and spruce are the most popular species of wood used for board sheathing. The proper methods of finishing the sheathing at the gable ends and eaves of a house are detailed in the next chapter.

PROGRESS CHECK

1. The main purpose of roof sheathing is to _____ and _____ the structure.

2. The most popular roof sheathing material being used today is _____.

3. What is the minimum thickness of plywood recommended for roof sheathing?

4. Roof sheathing is started at the ridge and brought down to the rafter tails. T — F.

5. Plywood sheathing is placed so that the face grain is parallel to the roof rafters. T — F.

6. It is not necessary to leave any space between the edges and ends of plywood sheathing. T — F.

7. All vertical joints must be _____.

8. Nails are spaced _____ inches apart along the perimeter and _____ inches apart in the field of plywood sheathing.

9. For 1/2-inch plywood, _____ penny nails are used.

10. For 1/2-inch to 1-inch plywood, _____ penny nails are used.

11. Plywood sheathing should be covered with building felt to protect it from weather until the roofing material is applied. T — F.

12. Lumber sheathing is usually used when _____ _____ or _____ _____ is the finish roofing material.

13. One-inch nominal thickness by 4-, 6-, or 8-inch widths are the best sizes for board sheathing. T — F.

14. When board sheathing is installed with spaces between the edges, it is called _____ sheathing.

15. When board sheathing is installed with no space between the edges, it is called _____ sheathing.

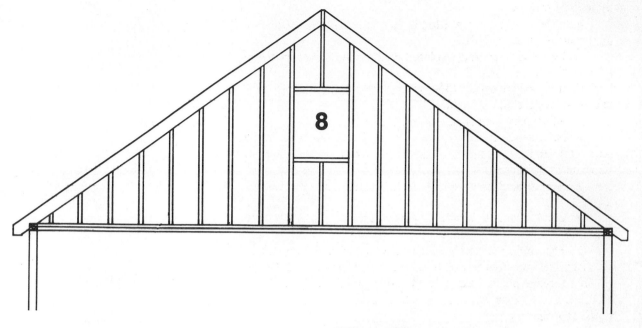

The Shed Roof

Objective. The shed roof is a popular and easy roof to erect. It is often used in conjunction with a gable or any other style roof to expand an attic. When used for this purpose it is built as a full dormer—usually across to full or nearly full length of the house— or built as a small dormer for light or ventilation. Upon completion of this chapter, you should understand and be able to do the following.

☐ Frame a roof opening.
☐ Calculate a shed rafter.
☐ Lay out and cut a shed rafter.
☐ Erect a shed roof.

The shed roof is one half of a gable or double-pitched roof. The pitch on a shed roof can be very small (almost flat) or it can be fairly steep, depending on its design. A shed roof can be used as a freestanding structure (Fig. 8-1). It can be used as a roof on an addition to an existing building, joining the building at some point on a vertical wall (Fig. 8-2), or it can intersect the building at some point on the roof if it is a gable or pitched roof (Fig. 8-3).

Where the use of clerestory windows are desired, a shed style roof is always used (Fig. 8-4). When you want to expand the attic in a house that has been framed using conventional style construction, a dormer with a shed roof is usually chosen. It gives the maximum amount of additional space, blends in nicely with most style

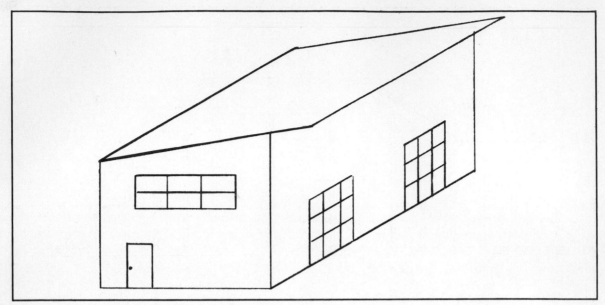

Fig. 8-1. Shed roof on a building that stands by itself.

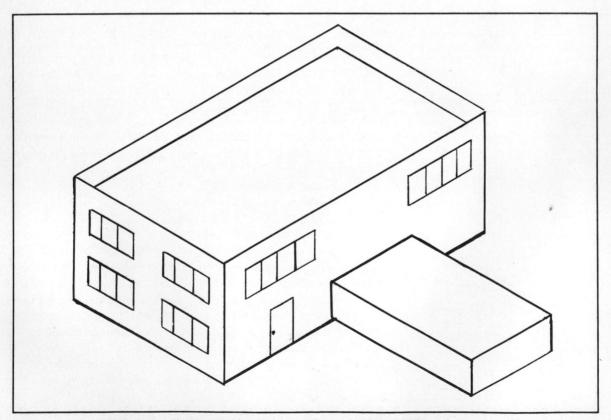

Fig. 8-2. A wing addition with a shed roof intersecting a building on a vertical wall.

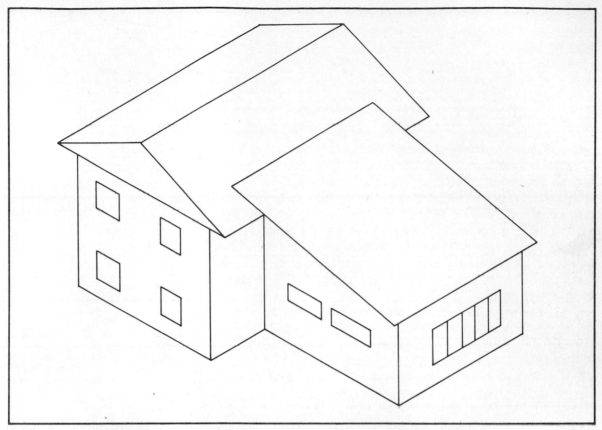

Fig. 8-3. A wing addition with a shed roof intersecting a building with a double-pitched roof.

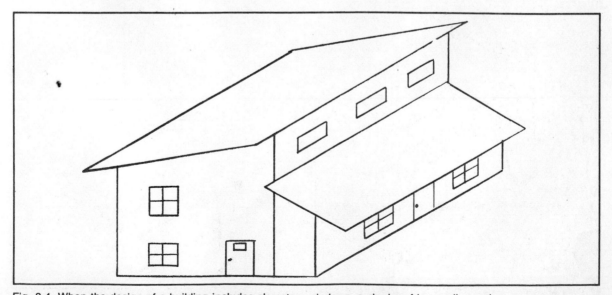

Fig. 8-4. When the design of a building includes clerestory windows, a shed roof is usually used.

roofs, and is comparatively easy to construct. A large dormer such as this can be constructed as the house is being built or added to an existing structure (Fig. 8-5). If it is for a single window or ventilation and covers only a portion of the roof, it is called a small dormer (Fig. 8-6).

THE FREESTANDING SHED ROOF

Figure 8-7 illustrates a section view of a shed roof that is constructed as a building that stands by itself. This type of roof usually has:

1. Two walls of different heights.

2. It consists of common rafters that rest on the two walls.

3. The rafter is laid out and calculated the same as a common rafter for a gable roof.

4. The shed rafter differs from a gable roof common rafter in the fact that it has two bird's-mouths one at each end of the rafter.

5. The heel cuts of the bird's-mouths are laid out toward the same end. The two heel plumb lines are marked in the direction of the lower wall end.

6. The run of the rafter is from the outside of the lower wall to the inside of the higher wall.

7. The total rise of the roof is the actual difference in height between the two walls.

8. The total rise can also be calculated by multiplying the unit rise by the run (the same as for a gable roof).

Fig. 8-5. A dormer covering a large portion of a house under construction.

148

Fig. 8-6. A small dormer with a shed roof constructed on a gable roof.

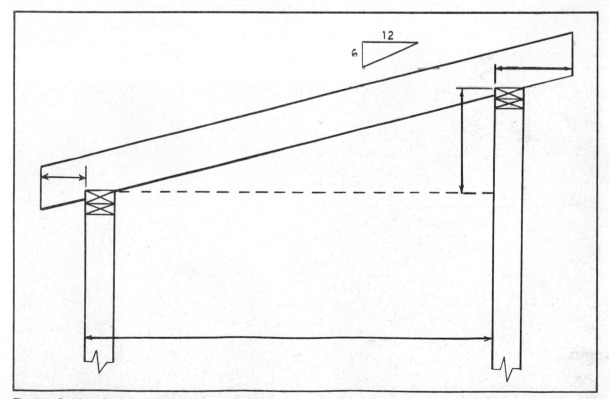

Fig. 8-7. Section view of a freestanding shed roof.

9. The run for the overhang is its projection. Notice how the projections are measured. They are from the outside edge of the plates to the tail cut on the lower wall and from the inside edge of the plates to the tail cut on the higher wall.

Laying Out a Rafter for a Freestanding Shed Building

The section plan of a shed roof type building supplies the following information shown in Fig. 8-8.

    a. The width of the building is 12' 0".

    b. The slope of the roof or cut is 6 and 12 (13.42 per ft. run).

    c. The wall sheathing thickness is 1/2 inch.

    d. The studs are 3 1/2 inch actual thickness.

    e. Combining sheathing and stud thickness gives a total wall thickness of 4 inches.

    f. The run of the rafter is the width of the building 12 feet minus the total wall thickness of 4 inches, for a result of 11' 8".

    g. The rise of the roof is the run 11' 8" converted to feet and decimals of a foot (11.66 feet) times the unit rise (6 inches), for a result of 69.96 inches, converted to 5' 9 15/16".

Fig. 8-8. Section view of a layout for a shed rafter.

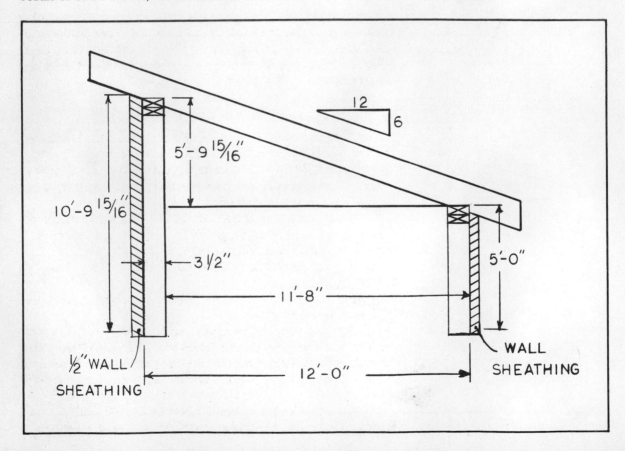

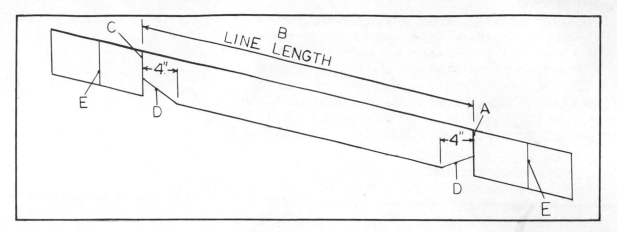

Fig. 8-9. Steps in laying out a shed rafter.

h. This results in the higher wall being 10' 9 15/16" with a lower wall of 5' 0".

i. Calculate the line length of the rafter by multiplying the run 11' 8" (11.66) by the per ft. run (13.42) for the unit rise of 6 inches. The result is 156.48 inches converted to 13' 0 1/2" for the length of the rafter.

j. Because sheathing thickness has to be included in the calculations, the seat cut of the bird's-mouth is cut at 4 inches on the shed rafter instead of 3 1/2 inches, as was used on the common rafter. The extra 1/2 inch is added in the direction of the tail cut from the heel plumb line.

Lay Out the Shed Rafter

As stated before, the shed rafter is laid out in the same fashion as the common rafter. The difference is that the shed rafter has two bird's-mouths (Fig. 8-9). Pick a good piece of rafter stock, crown it, and place the crown away from you. Start from the left and proceed as shown in the illustration.

a. At the given cut of the roof, draw the heel plumb line of the bird's-mouth that will be placed on the lower wall. Note that this line is plumb to the outside edge of the wall plates.

b. Step off the number of units run or measure the calculated line length and mark it.

c. Set the square to the cut of the roof at the line length mark and draw the heel cut line.

d. Set the square at right angles to the heel lines and draw in the seat cut lines for the bird's-mouths. If possible—and this depends on the depth of the rafter stock—the length of the seat cut should be the same dimension as the wall thickness (in this case 4 inches).

e. As done with the common rafter for a gable roof, slip the square to the desired projection length and mark off the overhang.

This is done at both ends of the shed rafter. Note that both projections do not have to be the same length; they can and sometimes do vary.

The shed rafters are then fastened to the walls of the building using the same nailing pattern that is used for the gable roof rafters. Gable studs are then calculated and placed using the same methods employed for the studs in the ends of a gable roof. Sheathing is applied as described in Chapter 7. Roof trim is applied as described in Chapter 9.

SHED DORMER

The shed dormer is a structure built onto a sloping roof. It is pitched in one direction and intersects the major roof somewhere on the slope. Shed dormers are classed as small or large. The small dormer is normally used for light and ventilation, and is sometimes used for just decoration. The large dormer spans a major portion of the house or sometimes the full length of the house. Large dormers expand the amount of attic living space by increasing the headroom to a standard 8 feet (or a little less, depending on the existing conditions). A shed dormer can be built on a new house as the roof is being framed, or it can be built onto an existing roof after an opening has been cut and framed in the roof.

Shown in Fig. 8-10 are the major parts of a small shed dormer. The dormer consists of the following framing members:

    a. Framed opening.

    b. End rafters doubled with a trimmer.

    c. Double headers placed plumb to the rafters.

    d. Front wall studs placed plumb, tops cut square, bottoms cut to the slope of the main roof.

    e. Gable studs spaced and placed plumb on the doubled-main roof common rafters. Bottoms cut to the slope of the main rafters. Tops cut to the slope of the shed roof.

    f. Cripple common rafters.

    g. Dormer rafters.

    h. Tail ends are cut plumb to desired projection.

Framing The Shed Dormer

Shown in Fig. 8-11 is the information needed to lay out and cut a rafter for a small shed dormer. Shown in detail D is the proper method of using the square to lay out the cut for the upper end of the rafter.

    a. To determine the run of the dormer rafter, divide the height of the end wall of the dormer, in inches, by the difference in the unit rise of main roof and unit rise of the dormer roof. With a unit of rise of 3 inches for the dormer roof and a unit of rise of 10 inches

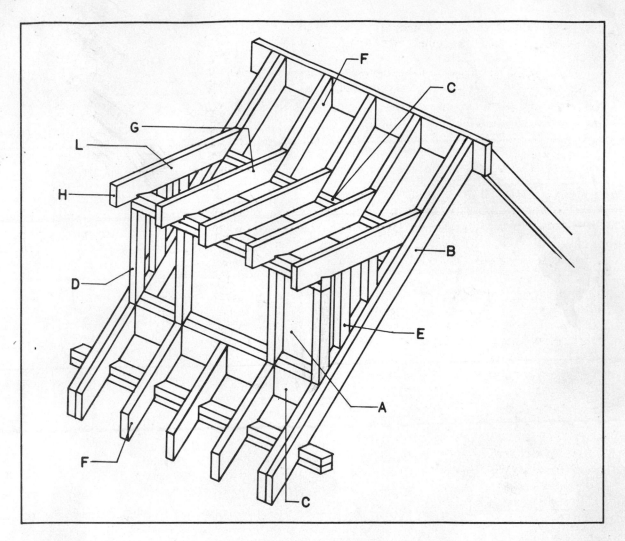

Fig. 8-10. Major parts of a small shed dormer.

for the main roof, the difference is 7 inches. The dormer wall is 8 feet multiplied by 12 inches, or 96 inches, divided by 7 inches. This gives an answer of 13.71 for the run of the dormer rafter.

b. Using the given unit of rise for the dormer rafter and calculated run, it is now possible to lay out the rafter with any method described for the common rafter in Chapter 6.

c. The bird's-mouth is cut using the unit of rise (3 inches) for the dormer rafter and the unit of run (12 inches).

d. The upper end of the rafter that intersects with the main roof is now cut. As shown in detail D, set the square to the cut of the main roof (10 and 12). Along the tongue of the square, measure down the unit of rise (3 inches) of the dormer rafter and mark it. Draw a line (shown by the dashed line) through this mark from the unit of run mark on the lower edge to the upper edge of the rafter.

153

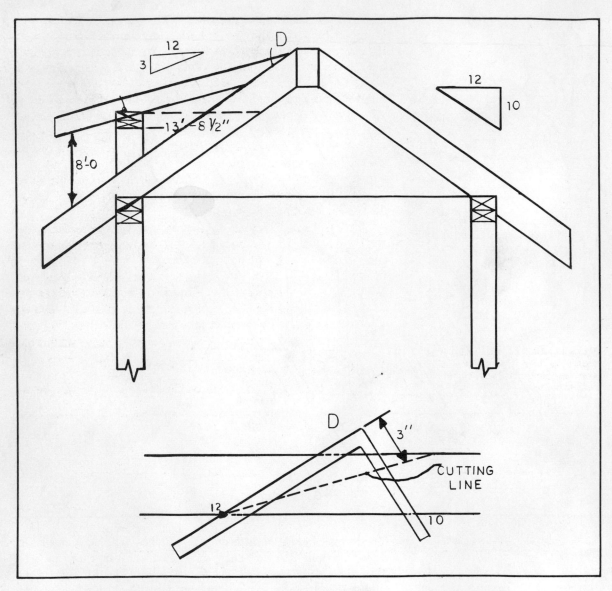

This is the cutting line that gives the proper angle for the correct fit.

The gable studs are measured and cut, the top cut is determined using the slope of the dormer rafter, and the bottom cut is determined using the cut of the main roof. The common difference in length of the gable studs is found by slipping the square (Fig. 8-12) the on center spacing of the gable studs, and reading the difference along the edge of the board.

Fig. 8-11. Layout and cuts of a small shed dormer.

A Large Dormer Shed Roof

Shown in Fig. 8-13 is a drawing of the framing members used in constructing a full or large dormer. The opening for this dormer is similar to a small dormer, but with some modification. Notice that the dormer rafters extend from the dormer wall to the ridge, but the end rafters tie onto the double common rafters. The ceiling joists in the dormer extend from the rafters back toward the ridge and usually to the rafters on the other side of the roof. These joists can rest on the front wall, but more likely they will be placed as shown in the drawing. The joists act as combined collar ties and ceiling joists. The exact placement of these joists depends on the amount of headroom desired.

a. Front wall studs are spaced and placed plumb on top plate of wall below. The tops and bottoms are cut square, and can be framed for windows if desired.

b. The common rafters on the roof, used to create the opening, are doubled if the dormer is less than the full length of the main roof.

c. The corner studs of the front wall are cut to the slope of the main-roof common rafter at the bottom. The tops are cut square.

d. The gable studs of the dormer are spaced 16 or 24 inches on center and placed plumb. Tops are cut to the slope of the dormer roof rafters, and bottoms are to the cut of the main roof rafters.

e. The tail and bird's-mouth of the dormer rafter are cut to the slope of the dormer roof.

f. The ends at the top of the middle dormer rafters are cut to

Fig. 8-12. Method of slipping the square to find the common difference in length of gable studs for a shed dormer.

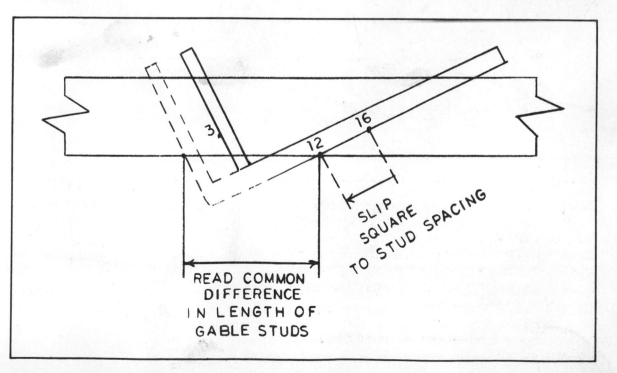

READ COMMON
DIFFERENCE
IN LENGTH OF
GABLE STUDS

SLIP
SQUARE
TO STUD SPACING

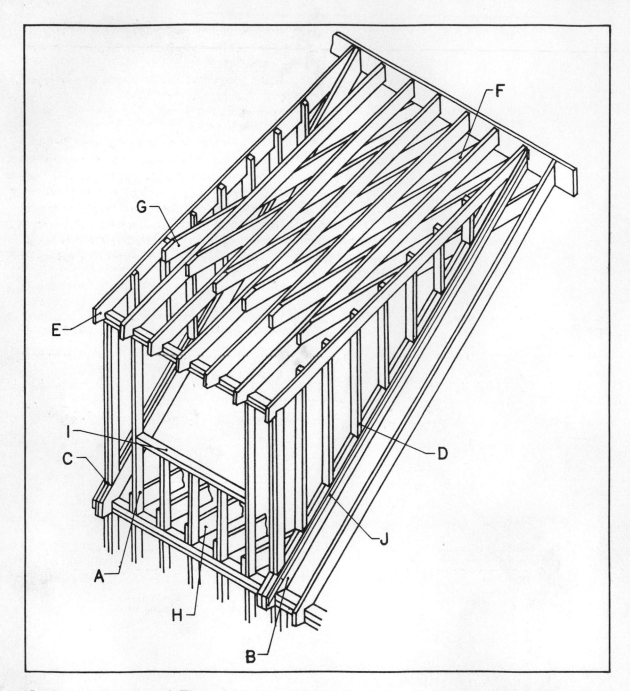

the slope of the main roof. The ends at the top of the end rafters are cut to the difference in slopes of the two roofs.

g. The ends of the dormer ceiling joists are cut to the slope of the dormer rafters and extend to the opposite side of the roof. The ends on that side are cut to the slope of the common rafter.

Fig. 8-13. Major framing members used in constructing a full or large dormer.

h. The partial joists are shown resting on the front wall plates and extend across the span of the house to form the attic floor and ceiling joists for the room below.

i. The front wall and rough opening for a vent or a window are framed the same as the main house walls.

j. The doubled common rafters have a ledger fastened to them to act as nailers for the roof sheathing.

Shown in Fig. 8-14 is a section view of a large shed dormer. This dormer can extend the full length of the house. Front wall studs are placed directly over and plumb to the exterior building wall below. The top cut of the rafter at the ridge is cut to the slope of the shed rafter and butts into the ridge. The heel plumb cut of the bird's-mouth is cut to the slope of the shed rafter. Make sure that the distance from the top edge of the shed rafter to seat cut (level line) of the bird's-mouth is the same as that of the common rafters in the rest of the roof.

Laying Out the Shed Rafter

Fig. 8-14. A section view of a large shed dormer.

An example in laying out a shed rafter is shown in Fig. 8-15.

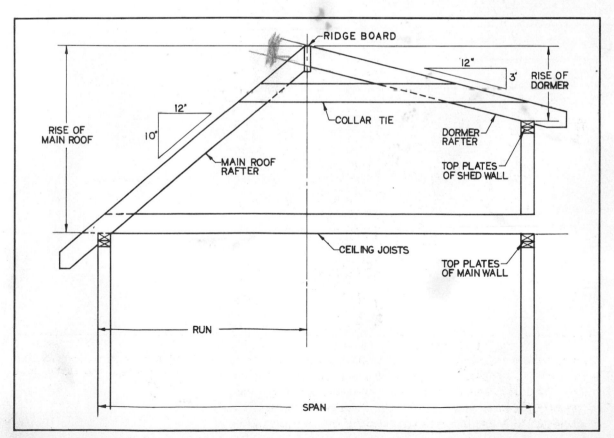

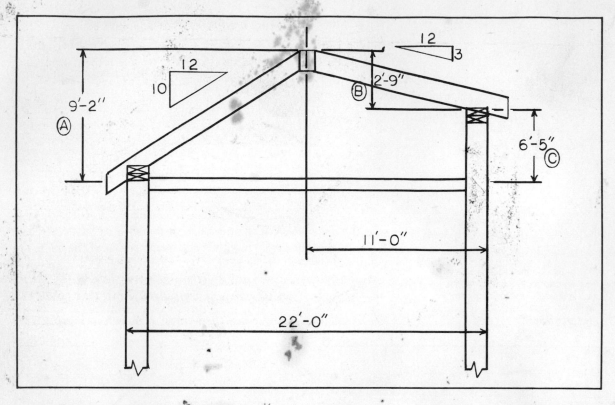

a. The rise of the main roof is determined by multiplying the unit rise of 10 inches by the run of 11 feet. Convert the result in inches, to feet for an answer of 9' 2".

b. The rise of the shed roof is calculated by multiplying the unit rise 3 inches by the run of 11 feet, for an answer of 33 inches converted to 2' 9".

c. The height of the shed roof wall is determined by finding the difference between the rise of the main roof and the rise of the shed roof. The result is the height of the shed wall, 6' 5 ".

d. The rafter line length can then be determined and marked by using either the step-off method or mathematically. If done using math, the answer is found to be 11' 4 1/16". This is accomplished by multiplying 12.37, the per ft. run, (for a unit rise of 3 inches) by the run of 11' 0". The result of 136.07 is then converted to feet and inches.

The top cut at the ridge, the bird's-mouth, the overhang, and the tail cut are all accomplished using the same methods described for the common gable rafter. Note that the deduction for ridge thickness must be made and that the distance down from the top edge of the rafter on the heel plumb line to the seat level line must be the same as the common rafter. The installation of the rafters and nailing pattern is the same as for any common rafter.

Fig. 8-15. Laying a shed rafter for a large dormer.

Framing a Dormer Roof Opening

When constructing a dormer, an opening must be built in the roof. This opening must be built to carry the weight of the dormer plus any loads placed on it. Described below are the parts and steps to frame an opening (Fig. 8-16).

The size and position of the dormer opening can be determined from the plans or drawings. A framing plan drawn to scale will be most helpful if made at this time. If accurately drawn, the location and dimensions of the framing members can be calculated. In actual construction, the framer will install all the main rafters, leaving out those rafters where the dormer opening will be framed. After the common rafters are placed, then the dormer opening and dormer are then framed.

Steps in Framing a Dormer Opening

a. Check the opening for the layout. Make sure doubled trimmers and cripple rafters are properly placed. They must be marked

Fig. 8-16. Illustrated are the parts and steps necessary to frame an opening in a roof for a dormer.

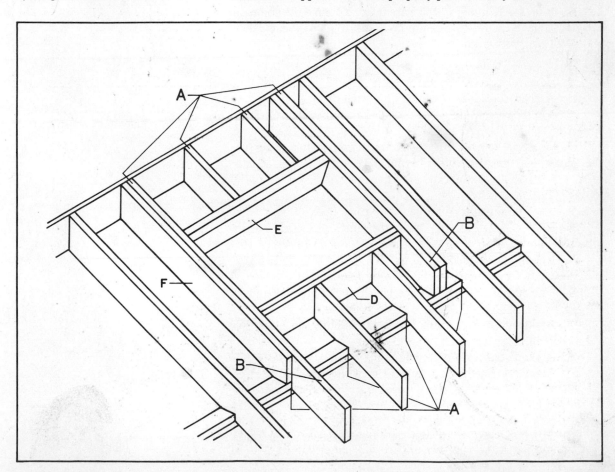

in the same position on the ridge board and the top plates of the building walls.

   b. Lay out and mark the inside rafters (both cripples and full length). The procedure for laying out and cutting the rafter is the same as that for the common rafter. There will be four full size rafters needed.

   c. Locate and lay out the upper and lower headers. Make sure the rough opening allows for doubling the headers. Place the header that is closest to the ridge and plates first. Don't forget to bevel the tops of the header to the same slope of the rafters.

   d. Lay out, cut, and install the cripple rafters. Don't forget to deduct for the header and ridge thickness. Face nail through the headers and ridge board into the cripple rafters. Toenail the tail ends to the top wall plates. Notice the headers are placed plumb to the rafters and the ends of the cripple rafters are cut to the given slope of the roof.

   e. Double the headers by nailing the inner ones to the outer headers; also nail them to the inner common roof rafter. It is not necessary for the trimmers, which are used to double the common rafters, to have tails. If needed to keep the 16 o.c. spacing, they should be cut with a tail.

PROGRESS CHECK

   1. A shed roof does not have any pitch. T — F.
   2. The bearing walls of a shed roof are different in height. T — F.
   3. The principle and methods of laying out a shed rafter are the same as those used in laying out a _____ _____.
   4. The shed rafter has _____ _____ whereas a common rafter in a gable roof has one.
   5. The _____ and the unit _____ must be known in order to lay out and cut a shed rafter.
   6. The thickness of the _____ _____ must be allowed for when cutting the bird's-mouth.
   7. The run of the rafter is the width of the building minus the _____ _____ thickness.
   8. A rafter for a freestanding shed building can have two overhangs of different sizes. T — F.
   9. A shed roof does not contain any gable studs. T — F.
   10. A shed dormer is usually built on a sloping roof. T — F.
   11. A small shed dormer is used for light or ventilation, whereas a large shed dormer increases the living area of a house. T — F.
   12. The front wall studs of a shed dormer have square cuts on both ends. T — F.

13. The gable studs on a sloping roof have angle cuts on both ends. T — F.

14. In framing a shed dormer, the run is determined by dividing the height of the dormer end wall, in inches, by the unit of rise. The answer will be in _____.

15. The upper end of a shed rafter, where it ties into the common rafters, must have a special cut. T — F.

16. The top cut on a gable stud is determined using the slope of the dormer roof. Whereas the bottom cut uses the slope of the main roof. T — F.

17. When the length of one gable stud is known, the difference in length of the remaining studs can be determined by _____ the square.

18. On a large dormer, the rafters extend from the dormer wall to the ridge of the main roof. T — F.

19. On a large dormer, the ceiling joists extend from the dormer to the other side of the house. T — F.

20. The ceiling joists on a shed dormer also act as _____ _____.

21. The end rafters on a large shed roof tie onto the doubled common rafter. T — F.

22. A ledger is fastened onto the doubled common rafter to act as a _____ for the _____.

23. The distance from the top edge of the shed roof rafter to the seat (level) cut of the bird's-mouth must be the same as the corresponding distance on the common rafter of the main roof. T — F.

24. Usually all the common rafters are installed, leaving space for the dormer opening which is built later. T — F.

25. The upper and lower headers in an opening must be _____ to the slope of the roof.

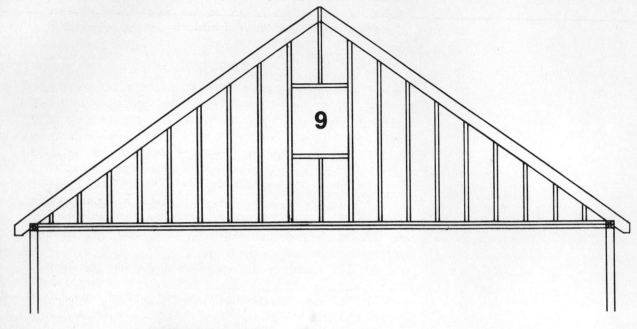

Roof Trim

Objective. Upon completion of this chapter, you should understand and be able to apply the following:
- ☐ The cornice.
- ☐ Fascia.
- ☐ Soffit.
- ☐ Rake.
- ☐ Materials used in trimming.
- ☐ Methods of trimming a roof.

With the rafters installed and the sheathing in place, the time has come to apply the "finishing touches" to the framed roof structure. This procedure is known as trimming. The schedule for trimming a house may vary depending on conditions such as the number of houses being built at the time, the structure of the builders crews, separate crews for the different phases of work, the progress payment schedule, and other factors.

Trim is installed for a number of purposes. It does finish exposed edges, it helps make weathertight joints, and it adds aesthetic value to the house. Some builders apply the fascia and rake trim, and then apply the roofing sheathing. Others apply the sheathing first and then the trim, followed by the roofing material. The choice of methods is not crucial. It is a matter of preference.

The trim should be completed before the finish roofing material is applied. Doing the job in this manner makes it much easier.

Today the trimming of a roof is kept relatively simple as compared to the methods using in the nineteenth and early twentieth century. Then the trimming of the roof was complex and very ornate.

CORNICES

The cornice is that portion of the roof that projects from the house. It is that part of the roof structure that covers and encloses the exposed rafter ends or overhang. A cornice can be constructed of metal, with an ornate design stamped in it, and applied to the eaves of the house. This type of construction was most popular in the past and used quite extensively in row house construction.

Wood, both plywood and boards, were and still are a popular material for cornice construction. Composition materials such as gypsum board, hardboard, and fiberboard are also used in the construction of a cornice. Combinations of wood, aluminum, and vinyl are also popular.

In cornice construction, the rafter tails continue past the wall plates to form an overhang. The distance the overhang projects out from the house is determined by the designer of the house. This distance is controlled by the height of the window heads and desired amount of protection from sun and weather.

There are three basic types of cornice construction used today:
☐ Close.
☐ Open.
☐ Box.

The Close Cornice

Figure 9-1 illustrates the construction of a close cornice.

a. The rafter ends are cut flush with the exterior walls.

b. The wall sheathing is brought up and butted with the roof sheathing.

c. The siding is installed up to the frieze board, which is placed even with the top edge of the wall sheathing.

d. A strip of moulding is used to finish off the trim. This is placed under the finish roof shingles. Note that the shingles hang over the moulding by a minimum of 3/4 of an inch.

Close cornice is not the most popular because it does not give the house protection from weather or sun, and does not lend itself to one of the most efficient methods of attic ventilation (soffit ventilation). Nevertheless, close cornice is the easiest and least costly cornice to construct.

A Narrow Box Cornice

Figure 9-2 illustrates a narrow box cornice.

a. The top edges of the rafters are covered with sheathing and

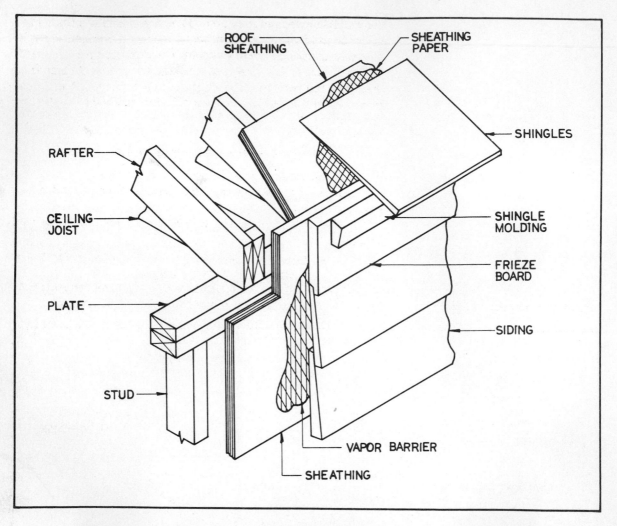

ROOF
SHEATHING

SHEATHING
PAPER

SHINGLES

RAFTER

CEILING
JOIST

SHINGLE
MOLDING

PLATE

FRIEZE
BOARD

SIDING

STUD

VAPOR BARRIER

SHEATHING

Fig. 9-1. Typical close cornice.

roof covering material over. The bottom edges of the rafters are covered with soffit material, and the front edges (tails) are covered with fascia material.

b. In the construction of a narrow soffit, the tail of the rafter receives a level cut onto which the soffit material is fastened.

c. Depending on the type of material used on the soffit and siding, it can be finished off with a frieze board and molding. Holes can be cut in the soffit an soffit vents installed for efficient attic ventilation.

The Box Cornice with Lookouts

The most common cornice used today is the box cornice with lookouts (Fig. 9-3). The construction of this cornice is similar to the narrow box cornice. Because this cornice can project out from the wall 24 inches and even up to 30 inches, a nailing surface must be supplied to which the soffit material can be fastened securely. This is done with the use of lookouts. The lookouts are usually made of 2-×-3-inch or 2-×-4-inch stock fastened in a level plane from the rafter tail back to the wall of the house. Screened vents can be placed in the soffit for a most effective method of ventilating the attic.

Box Cornice with a Sloping Soffit

Figure 9-4 illustrates a sloping soffit box cornice.

    a. In this type of cornice, the underside of the rafter does not have the level cut.

    b. The top edges of the rafters are covered with the roof sheathing and roof covering.

    c. The tails of the overhang are square cut. This places the fascia at an angle instead of being plumb.

    d. The underside of the rafters have no level return cut.

    e. The soffit material is fastened directly to the rafter.

Fig. 9-2. A narrow box cornice.

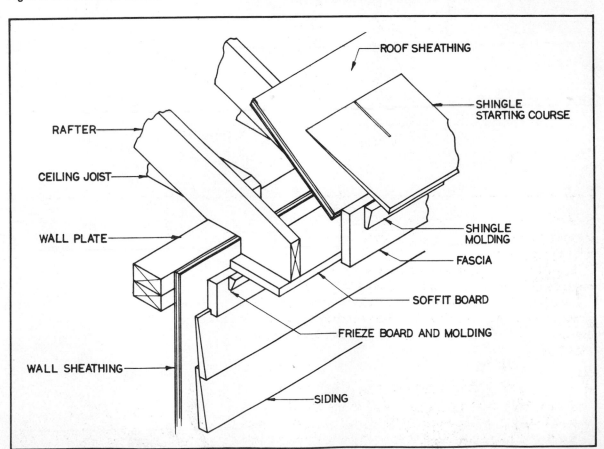

ROOF SHEATHING

SHINGLE STARTING COURSE

RAFTER

CEILING JOIST

WALL PLATE

SHINGLE MOLDING

FASCIA

SOFFIT BOARD

FRIEZE BOARD AND MOLDING

WALL SHEATHING

SIDING

165

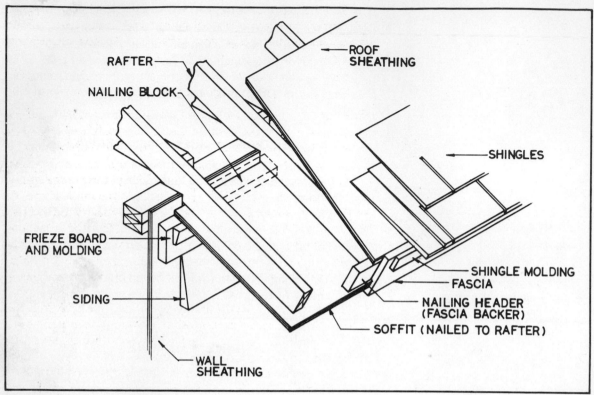

RAFTER

NAILING BLOCK

ROOF SHEATHING

SHINGLES

FRIEZE BOARD AND MOLDING

SIDING

SHINGLE MOLDING

FASCIA

NAILING HEADER (FASCIA BACKER)

SOFFIT (NAILED TO RAFTER)

WALL SHEATHING

Fig. 9-3. Box cornice with lookouts.

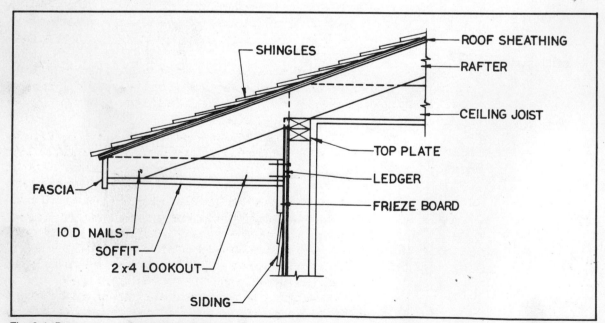

SHINGLES

ROOF SHEATHING

RAFTER

CEILING JOIST

TOP PLATE

LEDGER

FRIEZE BOARD

FASCIA

10 D NAILS

SOFFIT

2 x 4 LOOKOUT

SIDING

Fig. 9-4. Box cornice with a sloping soffit.

This method of construction creates a box cornice. Even though this type of cornice is relatively inexpensive to construct, it is not as popular as the narrow box cornice or the box cornice with lookouts.

TRIMMING THE GABLE END

Now that the eaves of the house have been finished with the construction of a cornice, the fascia boards are continued around the corner of the house and follow the slope of the roof to the ridge. This trim is referred to as the rake boards. They are also called barge boards or verge boards. The finishing of the gable ends of the house can vary from the very simple to the most ornate, depending on the style of the house. The continuation of the fascia boards around the corner and up the rake of the roof is the most common method used today. The trim boards are cut to the slope of the roof at the ridge and where it butts into the fascia. The plumb cut at these ends is obtained with the use of the square in the same manner as that used to lay out a common rafter.

Fascia and rake boards are usually white pine #2 B or better. The width will depend on the rafter stock; use of 1 × 6 or 1 × 8 is very common. At times, fascia will have to be ripped to size (depending on the design of the cornice).

Soffit can be selected from a number of materials. Plywood is the most popular. If plywood is used, choose exterior grade with the exposed side "A" grade. The hidden side can be "C" grade. If possible leave no exposed ends. It is best to use a minimum thickness of 3/8 inch. Any thinner material might cause a wave effect in the soffit. All joints, ends, and edges must have support above them.

Hardboard panels can also be used for soffit material. The installation of hardboard is the same as that for plywood. If possible, buy the panels with a primed surface. Otherwise be sure to prime the boards before installation.

Gypsum board is another material used for soffits. It can be used as long as it will not be exposed directly to the weather. Gypsum board is handled the same as Sheetrock. It is nailed with the same pattern, joints are taped and finished the same as Sheetrock, and exposed edges and edges that come in contact with other materials can be finished with metal trim. Gypsum can be cut by scoring and snapping the same as Sheetrock. Gypsum board is provided in 1/2-inch-thick, 4-foot-wide, and 8-to-12 foot lengths.

Of the materials mentioned, plywood is your best choice by far. Metal and plastic soffit materials are also very popular.

In the construction of a cornice, any framing should be done with 8d and 10d galvanized nails. Trim material in which the nails will be exposed should be fastened with 6d and 8d galvanized casing nails. Galvanized nails should always be used on any exposed, exterior construction. All nails should be countersunk and nail holes should be filled before painting. This will prevent rust stains.

PROGRESS CHECK

1. The portion of the roof that projects from the walls and covers the rafters is called a _____.

2. Name the three basic types of cornice _____, _____, _____.

3. A close cornice does not give the wall protection from weather or sun. T — F.

4. A _____ _____ is fastened to the ends of the rafters at the eaves of the house.

5. The board in question 4 is returned around the ends of the house and follows the slope up to the ridge. This board is called the _____ or at times the _____ _____.

6. Soffit material is usually fastened to the bottom of the rafters. T — F.

7. If this soffit material is fastened directly to the bottom of the rafters, the cornice is said to have a _____ soffit.

8. If the soffit material is installed in a level line, nailers must be used from the ends of the rafters to the house. These nailers are called _____.

9. Pine is a common material used for fascia trim. T — F.

10. A most effective method of venting an attic is with the use of soffit vents. T — F.

11. One of the most popular soffit materials used today is _____.

12. Hardboard and _____ board are two other common soffit materials.

13. A minimum thickness of plywood is _____ of an inch. A minimum thickness of gypsum board is _____ inch.

14. Soffit and fascia can be covered with _____ or _____ for a long-lasting finish.

15. Nails for exterior trim should always be coated with _____ to prevent rusting. They should also be countersunk and the holes _____ before painting.

Appendix A

Answers to Practice
Problems and Progress Check

Chapter 1:   Preliminary Roof Framing Information

1. Wood
2. F
3. T
4. 1/4
5. scaffolding
6. ladder
7. nails, bolts
8. 20 penny
9. F
10. steel framing square
11. 7 1/4
12. disconnect
13. full speed
14. 1/8
15. T
16. T
17. voltage tester
18. 3
19. ground adapter
20. F

Materials

1. F

2. hardwood, softwood
3. T
4. dried, planed
5. 35%, 300%
6. 19%
7. MC
8. kiln, air
9. T
10. T
11. F
12. two, hardwood, softwood
13. National Bureau of Standards
14. T
15. lowest
16. construction
17. Yes
18. Plywood
19. Lumber sheathing boards
20. right
21. T
22. native softwood
23. 5
24. T
25. CDX

Chapter 2: Basic Math and the Electronic Calculator

Addition
a. 39,578
b. 8,070.92
c. 254.03
d. 442
e. 626,280

Subtraction
a. 61928
b. 26
c. 244,727.5
d. 129.7
e. 2,908.3

Multiplication
a. 4,125
b. 1,471,239
c. 309.8
d. 143.0625
e. 31,900

Division
a. 50
b. 44.876
c. 0.1
d. 257.9658
e. 211

Using memory
a. 104
b. 87
c. 86
d. 6
e. 53.5

Square root using memory
a. 7.1589
b. 7.399
c. 16.97
d. 29.44
e. 161.969

Chapter 3: The Steel Framing Square

1. 5 5/8 inches
2. 13.42 inches
3. 13 7/16 inches
4. 35.34 inches
5. 8 11/16 and 12
6. 26 inches
7. tongue, blade, heel
8. front of the blade
9. line 1 on blade
10. unit rise, unit run, hypotenuse
11. unit rise
12. the hypotenuse of a right triangle is equal to the square root of the sum of squares of its other two sides.
13. 6 15/16 and 12

Chapter 4: Roof Styles

1. protects the house from weather
2. shed, gable, flat, hip, gambrel, lean-to, clerestory, intersecting valley
3. no
4. sheathing
5. F
6. T
7. F
8. T
9. yes
10. yes

Chapter 5: Roof Framing Terms and Their Relationship

1. F
2. 12
3. rise
4. 24, twice
5. F
6. level
7. plumb
8. angle
9. fraction
10. T

Practice Problems

(A)
a. 45 inches
b. 96 inches
c. 93.15 or 93 1/8 inches
d. 129.20 or 129 3/16 inches

(B)
a. 5.33 or 5 5/16
b. 2.90 or 2 7/8
c. 4 inches
d. 4 inches

(C)
a. 1/4
b. 1/9
c. 1/3
d. 3/4

(D)
a. 4 inches
b. 3 inches
c. 12 inches
d. 8 inches

Chapter 6: The Common Rafter

1. 1/4 or 1/2
2. span, run, slope
3. line or measurement
4. T
5. heel, seat or level cut, heel plumb cut
6. T
7. tongue, body
8. unit rise by run
9. thickness, ridge
10. level
11. T
12. F
13. T
14. T
15. T
16. run 12'-4", cut 8", projection 6"
17. F
18. T
19. F
20. ridge board
21. brace
22. collar beam

Chapter 7: Roof Sheathing

1. cover, protect
2. plywood
3. 5/16 or 3/8
4. F
5. F
6. F
7. staggered
8. 6" and 12"
9. 16d
10. 8d
11. T
12. wood shakes, slate shingles
13. T

14. open

Chapter 8: Shed Roof

1. F
2. T
3. common rafter
4. two bird's-mouths
5. run, rise
6. wall sheathing
7. total wall
8. T
9. F
10. T
11. T
12. F
13. T
14. felt
15. T
16. T
17. slipping
18. T
19. T
20. collar tie
21. T
22. nailer, sheathing
23. T
24. T
25. beveled

Chapter 9: Roof Trim

1. cornice
2. close, open, box
3. T
4. fascia board
5. barge, verge board
6. T
7. sloping
8. lookouts
9. T
10. T
11. plywood
12. gypsum
13. 3/8
14. metal, plastic
15. galvanize, filled

Appendix B

Roof Framing Formulas

☐ To find total rise when unit rise and total run are given:

Total Rise = Unit Rise × Total Run

Unit rise in inches.
Total run in feet
Answer in inches

☐ To find unit rise when total run and total rise are given:

Unit Rise = Total Run ÷ Total Rise

Total rise in inches
Total run in feet
Answer in inches

☐ To find pitch when total rise and total span are given:

Pitch = Total Rise ÷ Total Span

Total rise in feet
Total span in feet
Answer (pitch) stated as a fraction

☐ To find the unit rise when the pitch and span are given:

Unit Rise = (Pitch × Span × 12) ÷ Run

Pitch as a fraction
Span is in feet
Run is in feet
Answer will be in inches

☐ To find the hypotenuse of a right triangle:

$$C = \sqrt{A(2) + B(2)}$$

C = hypotenuse (line length)
B = base (run)
A = altitude (rise)

☐ To find the common difference in length of gable studs:

Common Difference = Stud Spacing × Unit Rise Over Unit Run

$$\text{Common Difference} = \frac{\text{Spacing} \times \text{slope}}{\text{unit run}}$$

Spacing: o.c. of studs in inches.
Slope: Inches of rise per foot of run.
Unit Run: 12 inches.
Common Difference: Difference in length of gable studs in inches.

☐ To find the area of a rectangle:

Area = Length × Width

Appendix C

Grade Stamps, Classification
of Lumber, and Nail Chart

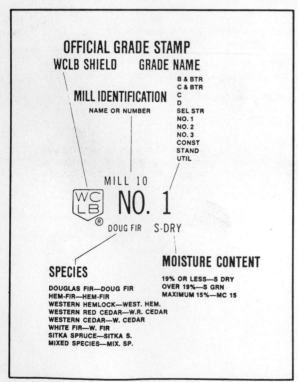

Fig. C-1. Explanation of markings on an official grade stamp. Courtesy of West Coast Lumber Inspection Bureau.

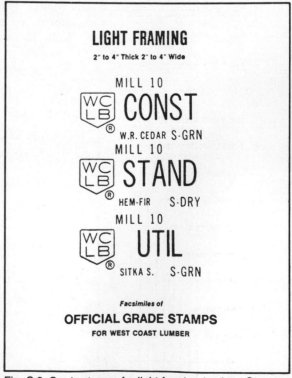

Fig. C-2. Grade stamps for light framing lumber. Courtesy of West Coast Lumber Inspection Bureau.

Grade Abbreviations

Grade	Abbreviation
B and Better	B&B
C and Better	C&Btr
Select Structural	Sel Str
Construction	Const
Standard	Stand
Utility	Util
Appearance	A
Air Seasoned	S-Dry
Kiln Dried	KD
Stress Rated	SR

Fig. C-3. Lumber grade abbreviations. Courtesy of West Coast Lumber Inspection Bureau.

SURFACED SIZES - IN INCHES

NOMINAL	GREEN	DRY
2	1-9/16	1-1/2
3	2-9/16	2-1/2
4	3-9/16	3-1/2
6	5-5/8	5-1/2
8	7-1/2	7-1/4
10	9-1/2	9-1/4
12	11-1/2	11-1/4
14	13-1/2	13-1/4
16	15-1/2	15-1/4

Fig. C-4. Surfaced sizes of construction lumber. Courtesy of West Coast Lumber Inspection Bureau.

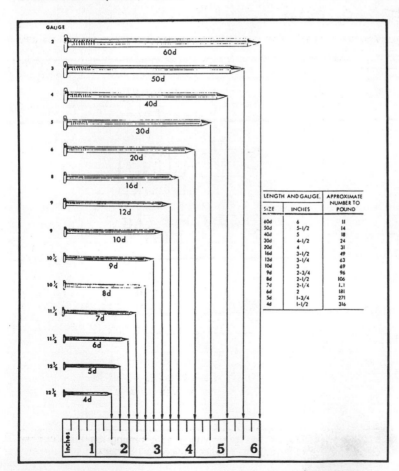

LENGTH AND GAUGE		APPROXIMATE NUMBER TO POUND
SIZE	INCHES	
60d	6	11
50d	5-1/2	14
40d	5	18
30d	4-1/2	24
20d	4	31
16d	3-1/2	49
12d	3-1/4	63
10d	3	69
9d	2-3/4	96
8d	2-1/2	106
7d	2-1/4	1.1
6d	2	181
5d	1-3/4	271
4d	1-1/2	316

Fig. C-5. Common nail size, gauge, length, and number per pound. Courtesy National Lumber Manufacturers Association.

175

Appendix D

Simplified Span Tables
for Light-Frame Roof Construction

RAFTER SPANS

SPANS ARE FOR RAFTERS USED IN COVERED STRUCTURES. SPANS APPLY TO LUMBER SURFACED "GREEN" OR SURFACED "DRY" WHICH CONFORMS TO PS 20 - 70 SIZES. APPLICABLE DESIGN CRITERIA ARE SHOWN IN THE HEADING FOR EACH TABLE. RAFTER SPANS ARE MEASURED ALONG THE HORIZONTAL PROJECTION (SEE FIGURE 1).

FIGURE 1 — Rafter Length / Rafter Span / HORIZONTAL PROJECTION

Spans are calculated using repetitive member values increased by 15% for two month duration of loading as for snow. Spans are shown in feet - inches.

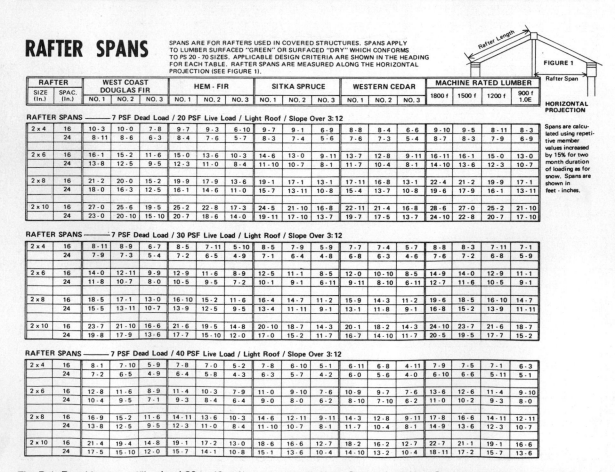

RAFTER SPANS — 7 PSF Dead Load / 20 PSF Live Load / Light Roof / Slope Over 3:12

RAFTER SIZE (In.)	SPAC. (In.)	WEST COAST DOUGLAS FIR			HEM-FIR			SITKA SPRUCE			WESTERN CEDAR			MACHINE RATED LUMBER			
		NO.1	NO.2	NO.3	NO.1	NO.2	NO.3	NO.1	NO.2	NO.3	NO.1	NO.2	NO.3	1800 f	1500 f	1200 f	900 f 1.0E
2 x 4	16	10-3	10-0	7-8	9-7	9-3	6-10	9-7	9-1	6-9	8-8	8-4	6-6	9-10	9-5	8-11	8-3
	24	8-11	8-6	6-3	8-4	7-6	5-7	8-3	7-4	5-6	7-6	7-3	5-4	8-7	8-3	7-9	6-9
2 x 6	16	16-1	15-2	11-6	15-0	13-6	10-3	14-6	13-0	9-11	13-7	12-8	9-11	16-11	16-1	15-0	13-0
	24	13-8	12-5	9-5	12-3	11-0	8-4	11-10	10-7	8-1	11-7	10-4	8-1	14-10	13-6	12-3	10-7
2 x 8	16	21-2	20-0	15-2	19-9	17-9	13-6	19-1	17-1	13-1	17-11	16-8	13-1	22-4	21-2	19-9	17-1
	24	18-0	16-3	12-5	16-1	14-6	11-0	15-7	13-11	10-8	15-4	13-7	10-8	19-6	17-9	16-1	13-11
2 x 10	16	27-0	25-6	19-5	25-2	22-8	17-3	24-5	21-10	16-8	22-11	21-4	16-8	28-6	27-0	25-2	21-10
	24	23-0	20-10	15-10	20-7	18-6	14-0	19-11	17-10	13-7	19-7	17-5	13-7	24-10	22-8	20-7	17-10

RAFTER SPANS — 7 PSF Dead Load / 30 PSF Live Load / Light Roof / Slope Over 3:12

RAFTER SIZE (In.)	SPAC. (In.)	WEST COAST DOUGLAS FIR			HEM-FIR			SITKA SPRUCE			WESTERN CEDAR			MACHINE RATED LUMBER			
		NO.1	NO.2	NO.3	NO.1	NO.2	NO.3	NO.1	NO.2	NO.3	NO.1	NO.2	NO.3	1800 f	1500 f	1200 f	900 f 1.0E
2 x 4	16	8-11	8-9	6-7	8-5	7-11	5-10	8-5	7-9	5-9	7-7	7-4	5-7	8-8	8-3	7-11	7-1
	24	7-9	7-3	5-4	7-2	6-5	4-9	7-1	6-4	4-8	6-8	6-3	4-6	7-6	7-2	6-8	5-9
2 x 6	16	14-0	12-11	9-9	12-9	11-6	8-9	12-5	11-1	8-5	12-0	10-10	8-5	14-9	14-0	12-9	11-1
	24	11-8	10-7	8-0	10-5	9-5	7-2	10-1	9-1	6-11	9-11	8-10	6-11	12-7	11-6	10-5	9-1
2 x 8	16	18-5	17-1	13-0	16-10	15-2	11-6	16-4	14-7	11-2	15-9	14-3	11-2	19-6	18-5	16-10	14-7
	24	15-5	13-11	10-7	13-9	12-5	9-5	13-4	11-11	9-1	13-1	11-8	9-1	16-8	15-2	13-9	11-11
2 x 10	16	23-7	21-10	16-6	21-6	19-5	14-8	20-10	18-7	14-3	20-1	18-2	14-3	24-10	23-7	21-6	18-7
	24	19-8	17-9	13-6	17-7	15-10	12-0	17-0	15-2	11-7	16-7	14-10	11-7	20-5	19-5	17-7	15-2

RAFTER SPANS — 7 PSF Dead Load / 40 PSF Live Load / Light Roof / Slope Over 3:12

RAFTER SIZE (In.)	SPAC. (In.)	WEST COAST DOUGLAS FIR			HEM-FIR			SITKA SPRUCE			WESTERN CEDAR			MACHINE RATED LUMBER			
		NO.1	NO.2	NO.3	NO.1	NO.2	NO.3	NO.1	NO.2	NO.3	NO.1	NO.2	NO.3	1800 f	1500 f	1200 f	900 f 1.0E
2 x 4	16	8-1	7-10	5-9	7-8	7-0	5-2	7-8	6-10	5-1	6-11	6-8	4-11	7-9	7-5	7-1	6-3
	24	7-2	6-5	4-9	6-4	5-8	4-3	6-3	5-7	4-2	6-0	5-6	4-0	6-10	6-6	5-11	5-1
2 x 6	16	12-8	11-6	8-9	11-4	10-3	7-9	11-0	9-10	7-6	10-9	9-7	7-6	13-6	12-6	11-4	9-10
	24	10-4	9-5	7-1	9-3	8-4	6-4	9-0	8-0	6-2	8-10	7-10	6-2	11-0	10-2	9-3	8-0
2 x 8	16	16-9	15-2	11-6	14-11	13-6	10-3	14-6	12-11	9-11	14-3	12-8	9-11	17-8	16-6	14-11	12-11
	24	13-8	12-5	9-5	12-3	11-0	8-4	11-10	10-7	8-1	11-7	10-4	8-1	14-9	13-6	12-3	10-7
2 x 10	16	21-4	19-4	14-8	19-1	17-2	13-0	18-6	16-6	12-7	18-2	16-2	12-7	22-7	21-1	19-1	16-6
	24	17-5	15-10	12-0	15-7	14-1	10-8	15-1	13-6	10-4	14-10	13-2	10-4	18-11	17-2	15-7	13-6

Fig. D-1. Dead load 7 psf/live load 20 to 40 psf/over 3 and 12 slope. Courtesy of West Coast Lumber Inspection Bureau.

RAFTER SPANS

SPANS ARE FOR RAFTERS USED IN COVERED STRUCTURES. SPANS APPLY TO LUMBER SURFACED "GREEN" OR SURFACED "DRY" WHICH CONFORMS TO PS 20 - 70 SIZES. APPLICABLE DESIGN CRITERIA ARE SHOWN IN THE HEADING FOR EACH TABLE. RAFTER SPANS ARE MEASURED ALONG THE HORIZONTAL PROJECTION (SEE FIGURE 1).

FIGURE 1

HORIZONTAL PROJECTION

Spans are calculated using repetitive member values increased by 15% for two month duration of loading as for snow. Spans are shown in feet - inches.

RAFTER SPANS — 10 PSF Dead Load / 20 PSF Live Load / No Finish Ceiling / 3:12 or Less Slope

RAFTER SIZE (In.)	SPAC. (In.)	WEST COAST DOUGLAS FIR			HEM - FIR			SITKA SPRUCE			WESTERN CEDAR			MACHINE RATED LUMBER			
		NO. 1	NO. 2	NO. 3	NO. 1	NO. 2	NO. 3	NO. 1	NO. 2	NO. 3	NO. 1	NO. 2	NO. 3	1800 f 2.1E	1500 f 1.8E	1200 f 1.5E	900 f 1.0E
2 x 6	16	14-8	14-4	10-11	13-9	12-10	9-10	13-9	12-4	9-5	12-4	12-0	9-5	15-5	14-8	13-9	12-4
	24	12-9	11-9	8-11	11-7	10-5	8-1	11-3	10-0	7-8	10-10	9-9	7-8	13-5	12-9	11-7	10-0
2 x 8	16	19-4	18-11	14-5	18-2	16-10	13-1	18-2	16-3	12-5	16-3	15-10	12-5	20-4	19-4	18-2	16-3
	24	16-9	15-6	11-9	15-3	13-9	10-8	14-10	13-3	10-1	14-4	12-11	10-1	17-8	16-9	15-3	13-3
2 x 10	16	24-7	24-1	18-5	23-2	21-6	16-8	23-2	20-8	15-10	20-8	20-3	15-10	25-11	24-7	23-2	20-8
	24	21-5	19-9	15-0	19-6	17-7	13-7	18-11	16-11	12-11	18-3	16-6	12-11	22-7	21-5	19-6	16-11
2 x 12	16	29-11	29-4	22-4	28-2	26-2	20-3	28-2	25-2	19-3	25-2	24-7	19-3	31-6	29-11	28-2	25-2
	24	26-0	24-1	18-3	23-9	21-5	16-6	23-0	20-6	15-8	22-10	20-1	15-8	27-6	26-0	23-9	20-6

RAFTER SPANS — 10 PSF Dead Load / 30 PSF Live Load / No Finish Ceiling / 3:12 or Less Slope

RAFTER SIZE (In.)	SPAC. (In.)	NO. 1	NO. 2	NO. 3	NO. 1	NO. 2	NO. 3	NO. 1	NO. 2	NO. 3	NO. 1	NO. 2	NO. 3	1800 f 2.1E	1500 f 1.8E	1200 f 1.5E	900 f 1.0E
2 x 6	16	12-9	12-6	9-6	12-1	11-1	8-7	11-11	10-8	8-2	10-9	10-5	8-2	13-5	12-9	12-1	10-8
	24	11-1	10-2	7-8	10-0	9-1	7-0	9-9	8-8	6-8	9-5	8-6	6-8	11-9	11-1	10-0	8-8
2 x 8	16	16-10	16-6	12-6	15-10	14-7	11-4	15-8	14-0	10-9	14-2	13-8	10-9	17-8	16-10	15-10	14-0
	24	14-8	13-5	10-2	13-3	11-11	9-2	12-10	11-6	8-9	12-5	11-3	8-9	15-6	14-7	13-3	11-6
2 x 10	16	21-6	21-0	15-11	20-3	18-8	14-6	20-0	17-11	13-8	18-2	17-6	13-8	22-6	21-6	20-3	17-11
	24	18-9	17-2	13-0	16-11	15-3	11-9	16-4	14-8	11-2	15-10	14-4	11-2	19-10	18-8	16-11	14-8
2 x 12	16	26-1	25-6	19-4	24-7	22-8	17-6	24-4	21-9	16-8	22-0	21-3	16-8	27-5	26-1	24-7	21-9
	24	22-9	20-10	15-10	20-6	18-6	14-4	19-11	17-9	13-7	19-3	17-4	13-7	24-1	22-8	20-6	17-9

RAFTER SPANS — 10 PSF Dead Load / 40 PSF Live Load / No Finish Ceiling / 3:12 or Less Slope

RAFTER SIZE (In.)	SPAC. (In.)	NO. 1	NO. 2	NO. 3	NO. 1	NO. 2	NO. 3	NO. 1	NO. 2	NO. 3	NO. 1	NO. 2	NO. 3	1800 f 2.1E	1500 f 1.8E	1200 f 1.5E	900 f 1.0E
2 x 6	16	11-7	11-2	8-5	10-10	9-11	7-8	10-8	9-6	7-3	9-11	9-3	7-3	12-3	11-7	10-10	9-6
	24	10-0	9-1	6-11	9-0	8-1	6-3	8-8	7-9	5-11	8-7	7-7	5-11	10-7	9-10	9-0	7-9
2 x 8	16	15-4	14-8	11-2	14-4	13-1	10-2	14-0	12-7	9-7	13-0	12-3	9-7	16-1	15-4	14-4	12-7
	24	13-3	12-0	9-1	11-10	10-8	8-3	11-6	10-3	7-10	11-3	10-0	7-10	14-1	13-1	11-10	10-3
2 x 10	16	19-6	18-9	14-3	18-4	16-8	12-11	17-11	16-0	12-3	16-7	15-8	12-3	20-6	19-6	18-4	16-0
	24	16-11	15-4	11-7	15-1	13-7	10-6	14-8	13-1	10-0	14-5	12-9	10-0	18-0	16-8	15-1	13-1
2 x 12	16	23-8	22-10	17-4	22-4	20-3	15-8	21-9	19-6	14-11	20-2	19-1	14-11	25-0	23-8	22-4	19-6
	24	20-6	18-7	14-1	18-4	16-7	12-10	17-9	15-11	12-2	17-5	15-7	12-2	21-10	20-3	18-4	15-11

Fig. D-2. Dead load 10 psf/live load 20 to 40 psf/no finish ceiling/3 and 12 slope or less. Courtesy of West Coast Lumber Inspection Bureau.

Desired angle	Use numbers	Decimal equivalent
15°	12 and 3 3/16	3.21
20°	12 and 4 3/8	4.37
25°	12 and 5 5/8	5.60
30°	12 and 6 15/16	6.92
35°	12 and 8 3/8	8.40
40°	12 and 10 1/16	10.07
45°	12 and 12	12.00

Fig. D-3. Figures to use on a framing square to obtain a given angle, the number 12 inches is in a horizontal plane, the variable number in inches is in a vertical plane, the decimal equivalent's are in inches.

RAFTER SPANS

SPANS ARE FOR RAFTERS USED IN COVERED STRUCTURES. SPANS APPLY TO LUMBER SURFACED "GREEN" OR SURFACED "DRY" WHICH CONFORMS TO PS 20 - 70 SIZES. APPLICABLE DESIGN CRITERIA ARE SHOWN IN THE HEADING FOR EACH TABLE. RAFTER SPANS ARE MEASURED ALONG THE HORIZONTAL PROJECTION (SEE FIGURE 1).

FIGURE 1

HORIZONTAL PROJECTION

RAFTER SPANS —— 15 PSF Dead Load / 20 PSF Live Load / Drywall Ceiling / All Slopes

RAFTER SIZE (In.)	SPAC. (In.)	WEST COAST DOUGLAS FIR			HEM - FIR			SITKA SPRUCE			WESTERN CEDAR			MACHINE RATED LUMBER			
		NO. 1	NO. 2	NO. 3	NO. 1	NO. 2	NO. 3	NO. 1	NO. 2	NO. 3	NO. 1	NO. 2	NO. 3	1800 f 2.1E	1500 f 1.8E	1200 f 1.5E	900 f 1.0E
2 x 6	16	14-6	13-4	10-1	13-2	11-10	9-2	12-9	11-5	8-8	12-6	11-2	8-8	15-4	14-6	13-2	11-5
	24	12-0	10-11	8-3	10-9	9-8	7-6	10-5	9-4	7-1	10-2	9-1	7-1	13-0	11-10	10-9	9-4
2 x 8	16	19-3	17-7	13-4	17-4	15-7	12-1	16-9	15-0	11-6	16-6	14-8	11-6	20-3	19-2	17-4	15-0
	24	15-10	14-4	10-10	14-2	12-9	9-10	13-8	12-3	9-4	13-5	12-0	9-4	17-1	15-7	14-2	12-3
2 x 10	16	24-7	22-5	17-0	22-1	19-11	15-5	21-5	19-2	14-8	21-0	18-9	14-8	25-10	24-5	22-1	19-2
	24	20-2	18-4	13-10	18-1	16-3	12-7	17-6	15-8	11-11	17-2	15-4	11-11	21-10	19-11	18-1	15-8
2 x 12	16	29-10	27-4	20-8	26-11	24-3	18-9	26-0	23-3	17-9	25-7	22-9	17-9	31-5	29-8	26-11	23-3
	24	24-7	22-3	16-11	21-11	19-10	15-3	21-3	19-0	14-6	20-10	18-7	14-6	26-7	24-3	21-11	19-0

RAFTER SPANS —— 15 PSF Dead Load / 30 PSF Live Load / Drywall Ceiling / All Slopes

RAFTER SIZE (In.)	SPAC. (In.)	WEST COAST DOUGLAS FIR			HEM - FIR			SITKA SPRUCE			WESTERN CEDAR			MACHINE RATED LUMBER			
		NO. 1	NO. 2	NO. 3	NO. 1	NO. 2	NO. 3	NO. 1	NO. 2	NO. 3	NO. 1	NO. 2	NO. 3	1800 f 2.1E	1500 f 1.8E	1200 f 1.5E	900 f 1.0E
2 x 6	16	12-9	11-9	8-11	11-7	10-5	8-1	11-3	10-0	7-8	11-0	9-9	7-8	13-4	12-9	11-7	10-0
	24	10-7	9-7	7-3	9-6	8-6	6-7	9-2	8-2	6-3	9-0	8-0	6-3	11-6	10-5	9-6	8-2
2 x 8	16	16-9	15-6	11-9	15-3	13-9	10-8	14-10	13-3	10-1	14-7	12-11	10-1	17-8	16-9	15-3	13-3
	24	13-11	12-8	9-7	12-6	11-3	8-9	12-1	10-10	8-3	11-10	10-7	8-3	15-1	13-9	12-6	10-10
2 x 10	16	21-4	19-9	15-0	19-6	17-7	13-7	18-11	16-11	12-11	18-7	16-6	12-11	22-7	21-4	19-6	16-11
	24	17-10	16-2	12-3	15-11	14-4	11-1	15-5	13-9	10-6	15-2	13-5	10-6	19-3	17-7	15-11	13-9
2 x 12	16	26-0	24-1	18-3	23-9	21-5	16-6	23-0	20-6	15-8	22-7	20-1	15-8	27-6	26-0	23-9	20-6
	24	21-8	19-8	14-11	19-4	17-5	13-6	18-9	16-9	12-10	18-5	16-5	12-10	23-6	21-4	19-4	16-9

RAFTER SPANS —— 15 PSF Dead Load / 40 PSF Live Load / Drywall Ceiling / All Slopes

RAFTER SIZE (In.)	SPAC. (In.)	WEST COAST DOUGLAS FIR			HEM - FIR			SITKA SPRUCE			WESTERN CEDAR			MACHINE RATED LUMBER			
		NO. 1	NO. 2	NO. 3	NO. 1	NO. 2	NO. 3	NO. 1	NO. 2	NO. 3	NO. 1	NO. 2	NO. 3	1800 f 2.1E	1500 f 1.8E	1200 f 1.5E	900 f 1.0E
2 x 6	16	11-7	10-8	8-0	10-6	9-5	7-4	10-2	9-1	6-11	10-0	8-10	6-11	12-2	11-7	10-6	9-1
	24	9-7	8-8	6-7	8-7	7-9	6-0	8-3	7-5	5-8	8-1	7-3	5-8	10-4	9-5	8-7	7-5
2 x 8	16	15-3	14-0	10-7	13-10	12-6	9-8	13-5	12-0	9-2	13-2	11-9	9-2	16-1	15-3	13-10	12-0
	24	12-7	11-5	8-8	11-3	10-2	7-11	10-11	9-9	7-6	10-9	9-6	7-6	13-8	12-5	11-3	9-9
2 x 10	16	19-6	17-11	13-7	17-8	15-11	12-4	17-1	15-3	11-8	16-9	14-11	11-8	20-6	19-6	17-8	15-3
	24	16-1	14-7	11-1	14-5	13-0	10-0	13-11	12-6	9-6	13-8	12-2	9-6	17-5	15-10	14-5	12-6
2 x 12	16	23-9	21-9	16-6	21-5	19-4	15-0	20-9	18-7	14-2	20-5	18-2	14-2	25-0	23-9	21-5	18-7
	24	19-7	17-9	13-6	17-6	15-9	12-3	17-0	15-2	11-7	16-8	14-10	11-7	21-2	19-4	17-6	15-2

Spans are calculated using repetitive member values increased by 15% for two month duration of loading as for snow. Spans are shown in feet - inches.

Fig. D-4. Dead load 15 psf/live load 20 to 40 psf/drywall ceiling/all slopes. Courtesy of West Coast Lumber Inspection Bureau.

Slope	Pitch	Angle
2 and 12	1/12	9° - 28'
3 and 12	1/8	14° - 2'
4 and 12	1/6	18° - 26'
6 and 12	1/4	26° - 34'
8 and 12	1/3	33° - 41'
10 and 12	5/12	39° - 46'

Fig. D-5. Three ways of stating the inclination of roof, the angle is measured in a horizontal plane.

Glossary

abut—Two ends that meet or touch. One structure that meets or joins another.

adhesive—A substance that holds materials together without mechanical fasteners. Mastics and glues are two common adhesives.

blocking—A small piece of wood placed across a joint to strengthen or stiffen it. Small blocks of wood placed between two structural members to stiffen them.

building codes—A set of standards and specifications for the construction industry to protect the health, welfare, and safety of the public. When adopted by a local municipal government, codes become legal documents.

butt joint—A joint made by fastening two members together, end to end.

clerestory—That portion of a wall that extends above the roof line. This allows for the installation of windows for light and ventilation.

collar beam—Also called collar tie, the collar beam connects two opposite rafters in a horizontal direction. It is usually placed in the upper third of the rafter and tends to make the roof stronger.

common difference—The difference in length between adjacent studs that intersect a rafter that is set at an angle. This difference in measurement will be the same for any studs that are evenly spaced along that rafter.

corner (inside)—The intersection of two exterior walls with the vertex of the angle facing away from the house.

corner (outside)—The intersection of two exterior walls with the vertex facing toward the house.

conventional roof framing—A roof that is built on the site using individual members as opposed to the use of factory-built trusses.

decking—The portion of a building that covers the skeleton of a roof or floor. It can be the standard board type, sheets of plywood, or a composition of other materials.

dormer—A framed opening that projects from a sloping roof to allow for light and ventilation.

eaves—The portion of a roof that extends or projects past the wall of a building.

elevation (drawing)—A drawing in a vertical plane showing the exterior walls as though they were being observed from straight ahead.

exterior walls—An outer wall enclosing a building; a wall that is completely exposed.

factory lumber—Graded lumber that is intended to be cut up for additional use. Also called shop lumber.

fascia—A flat horizontal member of a cornice placed on the outside ends of a rafter.

felt paper (building)—An asphalt-based material used on roofs as a protection against dampness.

foundation—The lowest part of a wall. The part of a wall, usually below ground, that supports the structure.

framing members—The skeleton parts of a building. On a roof they are rafters, on a floor they are joists, and on partitions they are the studs. Collectively they create the complete skeleton structure.

frieze board—A horizontal member that connects the top piece of siding with the soffit or cornice.

gable—The end of a building. The gable is associated with the triangular end of a building formed by the pitch of the roof.

galvanize—A coating of zinc applied to iron or steel to help prevent rusting. It is applied by hot dipping or electroplating.

girder—A large, horizontal member used to support walls or joists.

grading (lumber)—The process by which the quality and structural strength of lumber is determined for marketing.

grain—The arrangement of the fibers in a piece of wood. Working the wood in a longitudinal direction means working with the grain. Working in a transverse direction means going across the grain.

gypsum—A mineral, hydrous sulphate of calcium, processed and sandwiched between two sheets of special paper to form drywall. When fibers are added to the core it is called gypsum lath.

hardboard—A panel made out of pressed wood. Usually used on interior work but at times is used for soffit and siding.

inclined plane—The angle a surface makes with a horizontal line. The angle of inclination.

joists—A heavy, horizontal timber laid on edge that supports a floor or ceiling.

knee wall—partitions of varying height used to reduce the span on long rafter runs, thereby increasing the load-bearing capacity of the rafter.

load—In Carpentry, it is the weight of a structure or portion of a structure being supported by structural members below it, such as walls, columns or piers.

load (dead)—The weight of the structural members. Any fixed materials or parts of a structure.

load (live)—The variable load placed on a rafter or any other structural member. Snow or ice are the two greatest live loads placed on a pitched roof. Water can be a considerable load on a flat roof.

mansard (roof)—A roof with two slopes on all four sides. The lower slope is very steep and the upper slope is almost flat. Used to increase the living area of an attic. The hip roof differs from a mansard in that the hip roof only has one slope on all four sides.

member—A definite part of a building. Parts of a structure such as rafters, joists, studs, bracing.

nails—A fastener used in construction. A thin piece of pointed metal in a variety of shapes and sizes for specific uses.

nails (common)—Used when good holding power is required and appearance is not of any importance. Common nails are usually covered by some other material.

nails (finish)—Used when a nice appearance is important, finish nails have a small head that can be countersunk and the hole filled.

nail (casing)—Similar to a finish nail but with a slightly larger head, casing are usually used on exterior trim.

on center (o.c.)—The location of structural members. The center of one member to the center of another.

partition—A permanent interior wall separating the interior of a house into various rooms.

penny—Designates the size of nails. The system originated in England and indicates the price per hundred nails. Abbreviated d.

pitch of a roof—The angle a rafter makes with the wall plates and ridge of a roof. Expressed as a fraction.

plank—A long, flat piece of lumber usually 6 inches or more in width and 2 inches or more in thickness. Used in conjunction with scaffolding or horses to make temporary work platforms.

plate—A horizontal member placed on top of a foundation or stud wall to support the weight of the floor joist, ceiling joists, or roof rafters. Usually the same width as the walls. Can be either single or double.

primer—The first coat of paint applied to a surface. An undercoater.

projection—The jutting out of any part of a structure. In roof framing, the horizontal distance from the wall to the end of the rafter.

quarter-sawing—The sawing of a log into quarters, and then sawing the logs into boards.

rafter plate—The framing member on which the rafters sit. Also called the top plate.

rake—The inclined portion of a cornice. The angle that follows the slope of the roof. Used mostly in reference to the gable end construction.

roofing material—The covering used on the topmost part of a building; used to make the building watertight.

roof ties—Boards or planks fastened in the upper section of a roof to help keep the rafters and walls from spreading. Also called collar ties or collar beams.

sagging—The bending in the middle of a rafter, joist, beam, or any construction member due to its own weight or any load placed upon it.

sawyer—A person responsible for the sawing operations in a mill. That person who controls the cutting of logs into boards.

shakes—Hand split wood shingles.

sheathing—Plywood or wood boards that are placed over rafters of a roof to cover the structure.

span—The distance between the supporting members of a roof.

stud—An upright member of a wall partition.

timber—A growing tree. In construction referred to as a piece of lumber larger than 4 × 6 inches.

transverse—A line that lies across the length of the roof. It runs in the same direction as the rafters, and it is at right angles to the length of the roof.

unit length—Appears in the rafter table on the square. The hypotenuse of the unit triangle. The length 1 foot of rafter for every foot of run with for given rise.

upright—A member that stands in a vertical position.

valley—In roof framing, a depressed angle formed by the intersection of two inclined sides of the roof. As opposed to a hip which is a raised angle formed by two inclined sides.

wall (exterior)—An upright structure that encloses a building.

wall (interior)—An upright structure that divides the inside area of a building into usable living area.

weight-bearing wall—A wall that supports a load placed upon it in addition to its own weight. This type of wall can be an interior or exterior wall.

x (symbol)—Used in construction to indicate where to place a stud, joist, or rafter.

yard lumber—As defined in lumber grading rules, any lumber intended for general construction less than 5 inches in thickness.

zoning—As related to construction, laws that pertain to specific use of land.

Index

Other Bestsellers From TAB